Principles and Standards for School Mathematics Navigations Series

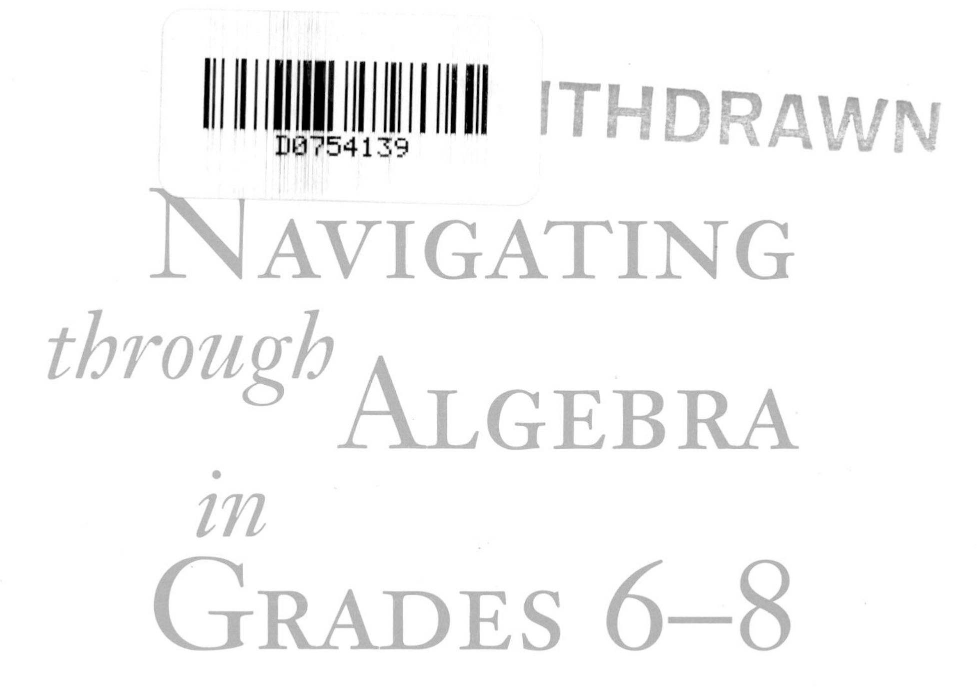

Navigating *through* Algebra *in* Grades 6–8

Susan Friel
Sid Rachlin
Dot Doyle

with

Claire Nygard
David Pugalee
Mark Ellis

Susan Friel
Grades 6–8 Editor

Peggy A. House
Navigations Series Editor

1906 Association Drive, Reston, VA 20191-1502
(703) 620-9840; (800) 235-7566; www.nctm.org

Fourth printing 2005

Library of Congress Cataloging-in-Publication Data:

Friel, Susan N.
Navigating through algebra in grades 6–8 / Susan Friel, Sid Rachlin, Dot Doyle ; with Claire Nygard, David Pugalee, Mark Ellis ; Susan Friel, grades 6–8 editor.
p. cm. — (Principles and standards for school mathematics navigations series)
Includes bibliographical references.
ISBN 0-87353-501-4
1. Algebra—Study and teaching (Middle school) I. Rachlin, Sid. II. Doyle, Dot. III. Title. IV. Series

QA159 .F75 2001
512′.071′2—dc21

2001030481

The National Council of Teachers of Mathematics is a public voice of mathematics education, providing vision, leadership, and professional development to support teachers in ensuring mathematics learning of the highest quality for all students.

Printed in the United States of America

Table of Contents

Readings from Publications of the National Council of Teachers of Mathematics

About This Book

In the discussion of the Algebra Standard in *Principles and Standards for School Mathematics* (National Council of Teachers of Mathematics [NCTM] 2000), NCTM takes the stand that by explicitly working to develop algebra concepts and algebraic thinking from the prekindergarten years, many, in fact most, students can complete a reasonable equivalent of algebra 1 by the end of grade 8. In the elementary grades, algebraic reasoning is developed informally. This initial development provides the background for a more systematic study of algebra in the middle grades. This book focuses on the development of algebraic reasoning and algebra concepts in grades 6–8. Two themes span the content of algebra at this level: (1) using mathematical models to represent and understand quantitative relationships and (2) representing and analyzing mathematical situations and structures. The concept of mathematical function, which encompasses both themes, receives considerable attention at these grade levels.

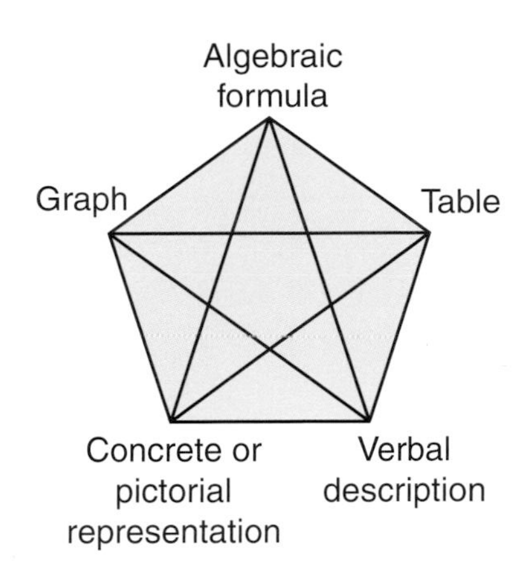

Explorations that develop from problems that can be solved by using tables, graphs, verbal descriptions, concrete or pictorial representations, or algebraic symbols offer opportunities for students to build their understandings of mathematical functions. The relationships among the five representations can be shown by the framework in the margin (Bowman 1993, p. 40).

The development of new vocabulary tied to algebraic concepts is important in the middle grades. Terms such as rate of change, variable, and continuous data label new ideas. A formal introduction of terminology is not initially as important as the modeling of its use by the teacher. As students gain experience with new ideas, they should be encouraged and expected to integrate the appropriate terms into their oral and written mathematical communication.

Each representation highlights some aspect of the concept of function, but no single representation can help students develop the deep understanding of functions that is needed. In fact, it is the processes connecting one representation to another that help students make sense of the concept of function. For example, moving from a table of data to a graph of those data may be done by constructing the graph by hand. The manual process involves such actions as drawing and scaling the x- and y-axes and plotting pairs of points. Alternatively, using a graphing calculator to move from the data to a graph involves choosing an appropriate graphing window and making sense of the axes as seen on the screen of the graphing calculator. Students may not be explicitly aware of the process of plotting points, even though they can identify specific points on the graph by using the trace key, for example. Students' understanding involves an awareness of the similarities and differences in the processes that occur when they construct a graph of tabular data by hand and with technology. Such an awareness helps them develop a richer understanding of the relationship between a table and a graph of a set of data—an important component in building an understanding of the concept of function.

The four chapters in this book consider some of the topics related to integrating the themes of using mathematical models and representing and analyzing mathematical situations and structures. The activities and problems require students to use multiple representations related to work with functions and highlight some of the interactions that may occur among these representations.

In chapter 1, "Understanding Patterns, Relations, and Functions," the topics include—

- clarifying the different uses of variables, and
- exploring, representing, analyzing, and generalizing a variety of

patterns using tables, graphs, words, and, when possible, symbolic rules.

Chapter 2, “Analyzing Change in Various Contexts,” includes—

- investigating how changes in one variable relate to changes in a second variable, and
- describing or portraying situations using graphs that may involve both constant and varying rates of change, with attention given to linear relationships.

In chapter 3, “Exploring Linear Relationships,” the topics include—

- identifying functions as linear or nonlinear and contrasting the properties of these two types of functions by examining tables, graphs, or equations;
- investigating relationships between symbolic expressions and the graphs of lines, paying particular attention to the meaning of intercept and slope; and
- using graphs to analyze the nature of changes in quantities in linear relationships.

Chapter 4, “Using Algebraic Symbols,” includes—

- using symbolic algebra to represent situations and to solve problems, especially those that involve linear relationships;
- recognizing and generating equivalent forms for simple algebraic expressions; and
- solving linear equations.

Each section in a chapter begins with a short discussion of important mathematical ideas. Then one or more strategies are presented that may provide insights into what students already know about those ideas. These strategies include examining the content of the curriculum from earlier in the year or previous years, as well as using preassessment tasks that may serve both to yield evidence of students' understanding and prior knowledge and to introduce a topic.

The remainder of each section presents one or more sample activities, many of which have blackline masters, which are signaled by an icon and can be found in the appendix. The CD-ROM, also signaled by an icon, contains two applets for students to manipulate and resources for professional development. There are many ways to help students develop the important mathematical ideas in the activities, but no matter how the curriculum is configured, it is important to pay attention to the mathematical content and to select activities that afford students an opportunity to a develop a conceptual understanding of important mathematics and, at the same time, to accomodate students' ways of making sense of information. Problems need to be sequenced carefully and taught according to a supportive instructional model that contributes to learning.

Although sequencing may be suggested by the order of the chapters, the interaction among the topics in these chapters is dynamic and is related to developing an increasingly deep and more sophisticated understanding of algebra. Thus, for example, although the study of patterns is used to stimulate the use of symbolic language and to introduce linearity, the topic may be revisited frequently, particularly as new families of functions are introduced and explored.

Key to Icons

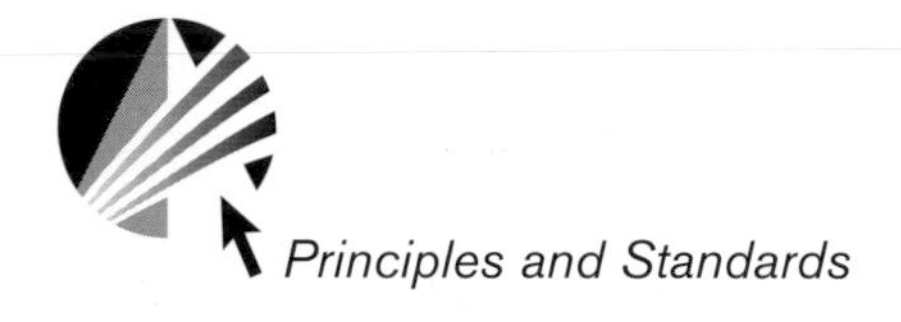
Principles and Standards

Blackline Master

CD-ROM

Three different icons appear in the book, as shown in the key. One alerts readers to material quoted from *Principles and Standards for School Mathematics,* another points them to supplementary materials on the CD-ROM that accompanies the book, and a third signals the blackline masters and indicates their locations in the appendix.

GRADES 6–8

NAVIGATING *through* ALGEBRA

Introduction

Throughout history, algebra has been a cornerstone of mathematics, and we can trace the roots of algebraic thinking deep into the bedrock of mathematics. Thus it is not surprising that algebra has emerged as one of the central themes of *Principles and Standards for School Mathematics* (National Council of Teachers of Mathematics [NCTM] 2000), for algebra continues to be an essential component of contemporary mathematics and its applications in many fields. Yet in the school curriculum, algebra has been too often misunderstood and misrepresented as an abstract and difficult subject to be taught only to a subset of secondary school students who aspire to study advanced mathematics; in truth, algebra and algebraic thinking are fundamental to the basic education of all students, beginning in the earliest years.

Algebra is frequently described as "generalized arithmetic," and indeed, algebraic thinking is a natural extension of arithmetical thinking. Both arithmetic and algebra are useful for describing important relationships in the world. But although arithmetic is effective in describing static pictures of the world, algebra is dynamic and a necessary vehicle for describing a changing world. Even young children can appreciate the significance of change and the need to describe and predict variation. Algebraic thinking begins with the very young, expands and deepens and matures through the years, and continues to serve adults long after the end of formal schooling. To achieve that outcome requires an algebra curriculum that is coherent and developmental, that is anchored by important mathematical concepts, and that is well articulated and coordinated across the grades.

The Navigations series seeks to guide readers through the five strands of *Principles and Standards for School Mathematics* in order to help them translate the Standards and Principles into action and to illustrate the growth and connectedness of content ideas from prekindergarten through grade 12. The Navigations through the algebra curriculum reflect NCTM's vision of how algebraic concepts should be introduced, how they grow, what to expect of students during and at the end of each grade band, how to assess what students know, and how selected instructional activities can contribute to learning.

Fundamental Components of Algebraic Thinking

The Algebra Standard emphasizes relationships among quantities and the ways in which quantities change relative to one another. To think algebraically, one must be able to understand patterns, relations, and functions; represent and analyze mathematical situations and structures using algebraic symbols; use mathematical models to represent and understand quantitative relationships; and analyze change in various contexts. Each of these basic components evolves as students grow and mature.

Understanding patterns, relations, and functions

Young children begin to explore patterns in the world around them through experiences with such things as color, size, shape, design, words, syllables, musical tones, rhythms, movements, and physical objects. They observe, describe, repeat, extend, compare, and create patterns; they sort, classify, and order objects according to various characteristics; they predict what comes next and identify missing elements in patterns; they learn to distinguish types of patterns, such as growing or repeating patterns.

In the higher elementary grades, children learn to represent patterns numerically, graphically, and symbolically, as well as verbally. They begin to look for relationships in numerical and geometric patterns and analyze how patterns grow or change. By using tables, charts, physical objects, and symbols, students make and explain generalizations about patterns and use relationships in patterns to make predictions.

Students in the middle grades explore patterns expressed in tables, graphs, words, or symbols, with an emphasis on patterns that exhibit linear relationships (constant rate of change). They learn to relate symbolic and graphical representations and develop an understanding of the significance of slope and y-intercept. They also explore "What if?" questions to investigate how patterns change, and they distinguish linear from nonlinear patterns.

In high school, students create and use tables, symbols, graphs, and verbal representations to generalize and analyze patterns, relations, and functions with increasing sophistication, and they convert flexibly among various representations. They compare and contrast situations modeled by different types of functions, and they develop an understanding of classes of functions, both linear and nonlinear, and their properties. Their understanding expands to include functions of more than one variable, and they learn to perform transformations such as composing and inverting commonly used functions.

Representing and analyzing mathematical situations and structures using algebraic symbols

Young children can illustrate mathematical properties (e.g., the commutativity of addition) with objects or specific numbers. They use objects, pictures, words, or symbols to represent mathematical ideas and relationships, including the relationship of equality, and to solve problems. When children are encouraged to describe and represent quantities in different ways, they learn to recognize equivalent representations and expand their ability to use symbols to communicate their ideas.

Later in the elementary grades, children investigate, represent, describe, and explain mathematical properties, and they begin to generalize relationships and to use them in computing with whole numbers. They develop notions of the idea and usefulness of variables, which they may express with a box, letter, or other symbol to signify the idea of a variable as a placeholder. They also learn to use variables to describe a rule that relates two quantities or to express relationships using equations.

During the middle grades, students encounter additional uses of variables as changing quantities in generalized patterns, formulas, identities, expressions of mathematical properties, equations, and inequalities. They explore notions of dependence and independence as variables change in relation to one another, and they develop facility in recognizing the equivalence of mathematical representations, which they can use to transform expressions; to solve problems; and to relate graphical, tabular, and symbolic representations. They also acquire greater facility with linear equations and demonstrate how the values of slope and y-intercept affect the line.

High school students continue to develop fluency with mathematical symbols and become proficient in operating on algebraic expressions in solving problems. Their facility with representation expands to include equations, inequalities, systems of equations, graphs, matrices, and functions, and they recognize and describe the advantages and disadvantages of various representations for a particular situation. Such facility with symbols and alternative representations enables them to analyze a mathematical situation, choose an appropriate model, select an appropriate solution method, and evaluate the plausibility of their solution.

Using mathematical models to represent and understand quantitative relationships

Very young children learn to use objects or pictures, and, eventually, symbols to enact stories or model situations that involve the addition or subtraction of whole numbers. As they progress into the upper elementary grades, children begin to realize that mathematics can be used to model numerical and geometric patterns, scientific experiments, and other physical situations, and they discover that mathematical models have the power to predict as well as to describe. As they employ graphs, tables, and equations to represent relationships and use their models to draw conclusions, students compare various models and investigate whether different models of a particular situation yield the same results.

Contextualized problems that can be modeled and solved using various representations, such as graphs, tables, and equations, engage middle-grades students. With the aid of technology, they learn to use functions to model patterns of change, including situations in which they generate and represent real data. Although the emphasis is on contexts that are modeled by linear relationships, students also explore examples of nonlinear relationships, and they use their models to develop and test conjectures.

High school students develop skill in identifying essential quantitative relationships in a situation and in determining the type of function with which to model the relationship. They use symbolic expressions to represent relationships arising from various contexts, including situations in which they generate and use data. Using their models, students conjecture about relationships, fomulate and test hypotheses, and draw conclusions about the situations being modeled.

Analyzing change in various contexts

From a very early age, children recognize examples of change in their environment and describe change in qualitative terms, such as getting taller, colder, darker, or heavier. By measuring and comparing quantities, children also learn to describe change quantitatively, such as in keeping track of variations in temperature or the growth of a classroom plant or pet. They learn that some changes are predictable but others are not and that often change can be described mathematically. Later in the elementary grades, children represent change in numerical, tabular, or graphical form, and they observe that patterns of change often involve more than one quantity, such as that the length of a spring increases as additional weights are hung from it. Students in the upper grades also begin to study differences in patterns of change and to compare changes that occur at a constant rate, such as the cost of buying various numbers of pencils at $0.20 each, with changes whose rates increase or decrease, such as the growth of a seedling.

Middle-grades students explore many examples of quantities that change and the graphs that represent those changes; they answer questions about the relationships represented in the graphs and learn to distinguish rate of change from accumulation (total amount of change). By varying parameters such as the rate of change, students can observe the corresponding changes in the graphs, equations, or tables of values of the relationships. High school students deepen this understanding of how mathematical quantities change and, in particular, of the concept of rate of change. They investigate numerous mathematical situations and real-world phenomena to analyze and make sense of changing relationships; interpret change and rates of change from graphical and numerical data; and use algebraic techniques and appropriate technology to develop and evaluate models of dynamic situations.

Developing an Algebra Curriculum

Clearly, an algebra curriculum that fosters the development of algebraic thinking described here and in *Principles and Standards* (NCTM 2000) must be coherent, focused, and well articulated. It cannot be merely a collection of lessons or activities but must instead be developmental and connected. Mathematical ideas introduced in the early years must deepen and expand, and subsequent instruction should link to, and build on, that foundation. As they progress through the curriculum, students must be continually challenged to learn and apply increasingly more-sophisticated algebraic thinking and to solve problems in a variety of school, home, and life settings.

The Navigations cannot detail a complete algebra curriculum, nor do they attempt to do so. Rather, the four Navigations through the Algebra Standard illustrate how a few selected "big ideas" of algebra should develop across the prekindergarten–grade 12 curriculum. Further, the topical strands of the mathematics curriculum, such as algebra, geometry, and data analysis, are highly interconnected, and many of the concepts presented under one strand will further develop and deepen when encountered again in another context. Thus, future Navigations through other parts of *Principles and Standards* (NCTM 2000) will reinforce the algebra objectives, and vice versa.

The methods and ideas of algebra are an indispensable component of mathematical literacy in contemporary life, and the algebra strand of the curriculum is central to the vision of mathematics education set forth in *Principles and Standards for School Mathematics.* These Navigations are offered as a guide to help educators set a course for successful implementation of the important Algebra Standard.

Grades 6–8

Navigating through Algebra

Chapter 1

Understanding Patterns, Relations, and Functions

In the recursive form of pattern generalization, students focus on the rate of change from one element to the next. For example, in the number sequence 7, 13, 19, 25, 31, 37, ..., the constant rate of change between successive terms is +6. To determine the tenth term using a recursive strategy means that a student would take the sixth term (37) and add 6 to determine the seventh term (43), then add 6 to determine the eighth term (49) and so on until the tenth term (61) is reached. The recursive form of pattern generalization in linear relationships highlights a constant rate of change.

Important Mathematical Ideas

Working with patterns is at the heart of mathematics. Generalizing involves perceiving some pattern or regularity and expressing it succinctly in order to communicate the perception and to use it to answer questions. Algebra is a tool for expressing such generalities (Mason, Graham, Pimm, and Gowar 1985). Exploring repeating patterns in the elementary grades lays the foundation for considering growing patterns (Van de Walle 2000) and number sequences in the middle grades. Exploring patterns is a vehicle to provoke thinking about variables and functions. Solution strategies for generalizing a pattern may be represented using tables, graphs, words, or algebraic symbols. Solution strategies may involve both *recursive* and *explicit* forms of pattern generalization. Various solution strategies and their expressions in words and with symbols naturally lead to a consideration of the equivalence of algebraic expressions.

Experiences in generating and describing growing sequences may be introduced through a variety of contexts, frequently geometric tasks. These tasks generally involve—

- finding the first few terms of the sequence and recording this information in a table or chart;
- drawing the next shape in the sequence;
- describing the shapes succinctly with words in such a way that someone who has not seen them will be able to duplicate the sequence;

In the explicit form of pattern generalization, the rule "Add 6" is related to the order of the terms in the sequence, so, for the first term, $7 = 1 + 6$; for the second term, $13 = 1 + 6 + 6$; for the third term, $19 = 1 + 6 + 6 + 6$, and so on. The general rule would be "term number $\times$ 6 + 1," or $6x + 1$. The explicit form of pattern generalization involves developing a rule or formula that highlights the relationship between the independent variable (the number representing the position of the term in the sequence) and the dependent variable (the number in the sequence).

- writing the rule that will produce the ever-growing sequence with the first few terms given (either a recursive or an explicit form of pattern generalization);
- generalizing to make predictions for the tenth, the forty-third, or the *n*th shape in the sequence;
- comparing different ways of arriving at the generalization (e.g., equivalence of algebraic expressions or recursive and explicit forms of pattern generalization).

What Might Students Already Know about These Ideas?

Given a curriculum as envisioned in *Principles and Standards for School Mathematics* (NCTM 2000), we expect that students will have had opportunities to explore informally the idea of function in grades 3–5, including opportunities to work with both repeating and growing patterns. The preassessment activity Exploring Houses may provide insights into how your students are thinking about variables, patterns, and functions as they enter the middle grades. The problem is similar to those encountered in grades 3–5.

Exploring Houses

Goal

To assess students'—

- understanding of pattern development;
- ability to use a table to organize information;
- choice of strategy (e.g., draw a picture, use recursion, write an explicit rule) to make a prediction.

Materials and Equipment

- A copy of the blackline master "Exploring Houses" for each student
- Enough pattern-block triangles and squares for each student to make the first four houses in the pattern

p. 74

Activity

The students should work individually. Have each student build the first four houses in the sequence in figure 1.1 as needed and then complete the following tasks:

1. For each house, determine the total number of pieces needed. How many squares and triangles are needed for a given house? Organize your information in some way.
2. Describe what house 5 would look like. Draw a sketch of this house.
3. Predict the total number of pieces you will need to build house 15. Explain your reasoning.
4. Write a rule that gives the total number of pieces needed to build any house in this sequence.

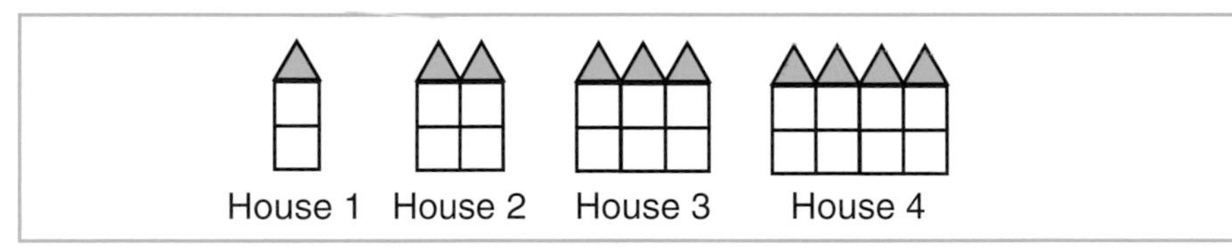

Fig. **1.1.**
A sequence of houses

Discussion

As the students explain how they determined the pattern, take note of their strategies. Do they build each pattern and notice the changes from one pattern to the next? Do they organize the information in a table? When they make predictions about other houses in the pattern, do they "add 3" multiple times (a recursive strategy) to predict the number of pieces needed or do they see a relationship between the house number and the number of pieces needed to make the house (an explicit strategy)? Do they talk about the total number of pieces, or do they talk about the total number of squares and the total number of triangles? Can they describe their rules in words? Using symbols? Answers to these questions as well as an overall analysis of students' written responses provide insights into how comfortable and capable students are in working with patterns and variables.

To solve this problem, students may not need to use a recursive strategy. They may see that each "house" in the sequence is built by adding a column of two squares and one triangle, or three additional pieces, to the previous house. So in the tenth house, there are 10×3 total pieces. The formula is $3n$, where n represents the house number.

A Table for the Houses Pattern

House Number	Total Number of Pieces
1	3
2	6
3	9
4	12
5	15
⋮	⋮
15	45

Selected Instructional Activities

Several examples of growing sequences are illustrated in figure 1.2; these patterns are drawn or made with various materials. Each pattern consists of a series of separate elements; each new element is related to the previous one on the basis of the pattern. It is important to engage middle school students in building patterns with physical materials such as tiles, toothpicks, counters, pattern blocks, and so on. This kind of pattern construction helps students focus on the physical changes and how the pattern is being developed.

Constructing a table or chart is a valuable strategy for organizing the numerical information that can be determined from the pattern. As students begin to look for relationships, some will focus on the numerical pattern in the table and others will attend to the physical pattern.

Fig. **1.2.**
Growing Patterns

1. Make growing patterns using the initials in your name. What is size 5 in the example shown?

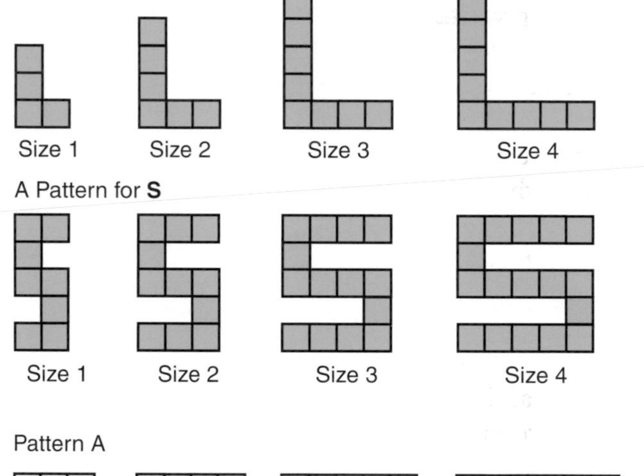

2. Make tiling patterns for garden areas. The tiles surround each garden. Two different patterns are shown. For each pattern, draw garden 5.

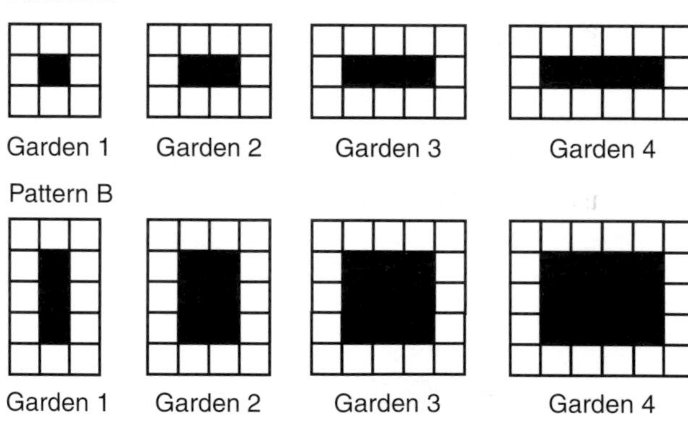

3. Use a one-unit stamp with paint on it. See how many stamps it takes to paint all the faces of each shape. Each rod in a shape is 2 units long. What happens to the number of stamps if each rod is 3 units long? 4 units long?

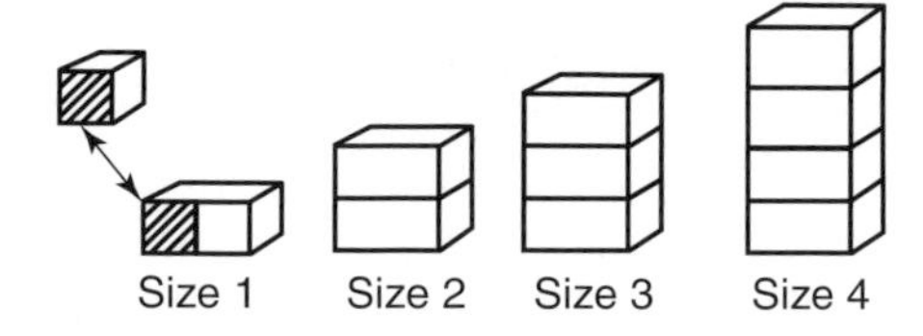

4. Use isometric dot paper to make shapes. How many unit line segments are in each shape shown? A unit line segment connects two adjacent dots.

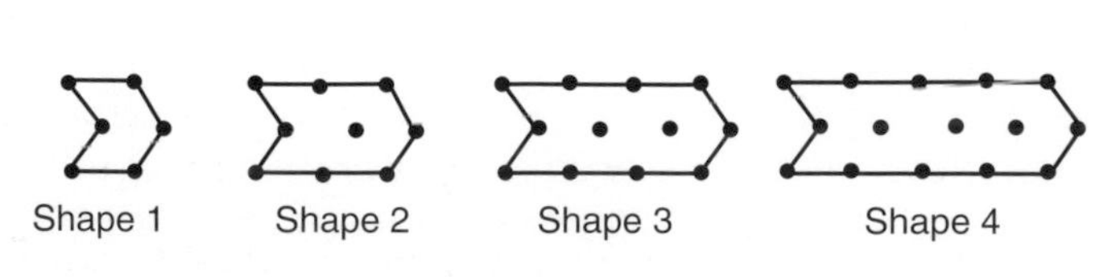

Challenging students to see how patterns observed in a table relate to the physical or pictorial model and vice versa helps them relate the two representations.

Observing how patterns change from one element to the next and then using a recursive strategy is often easier for students than defining an explicit rule for the pattern. For example, consider the first tiling pattern for garden areas in figure 1.2. The students might say, "The third garden increases by two tiles from the second garden." Their attention is on moving from one element to the next; in a table, the students are also looking at the change from one element to the next. They have identified a recursive relationship; if asked how many pieces are needed to make the twenty-fifth garden area, they would begin with the first garden area and continue to add 2 to get each successive garden area until they reached the twenty-fifth one. Students could also use the fill-down command in a spreadsheet program to generate the list of elements (see fig. 1.3). They can literally add 2 twenty-five times to help in identifying the twenty-fifth element by recursion. Given that students are working with a simple "+2" pattern, they should also recognize that they are adding forty-eight tiles (2×24) to the number of tiles found in the first element to get a total of fifty-six tiles.

Students can be encouraged to provide a justification for the "Add 2" recursive pattern they have observed. They can do so by examining what changes and what stays the same in each element of the sequence. So, in garden area 1, we see three columns: "3 tiles + 2 tiles + 3 tiles." In moving from garden area 1 to garden area 2, we see the number of two-tile columns change so that there are 3 tiles + 2 tiles + 2 tiles + 3 tiles. As we move to garden area 3, the number of two-tile columns changes again so that there are 3 tiles + 2 tiles + 2 tiles + 2 tiles + 3 tiles. A pattern emerges that can be used to identify the number of tiles needed for the twenty-fifth garden area:

Garden area 1 has eight tiles, or $3 + 2 + 3$ tiles.
Garden area 2 has ten tiles, or $3 + (2 + 2) + 3$ tiles.
Garden area 3 has twelve tiles, or $3 + (2 + 2 + 2) + 3$ tiles.
Garden area 25 has fifty-six tiles, or $3 + 25(2) + 3$ tiles.

An explicit rule, based on this pattern, might be stated for garden area n:

$$3 + n(2) + 3,$$

which is equivalent to $2n + 6$.

Other students may look at the gardens and notice that the widths are changing. In garden area 1, we see the two lengths of three tiles each and the remaining tile on each end of the garden. In garden area 2, the two lengths have four tiles each, but the single tile at either end remains the same.

Garden area 1 has eight tiles, or $1 + 3 + 3 + 1$ tiles.
Garden area 2 has ten tiles, or $1 + 4 + 4 + 1$ tiles.
Garden area 3 has twelve tiles, or $1 + 5 + 5 + 1$ tiles.
Garden area 25 has fifty-six tiles, or $1 + 27 + 27 + 1$ tiles.

An explicit rule, based on this pattern, might be stated for garden area n:

$$1 + (n + 2) + (n + 2) + 1,$$

One teacher noted that when her students made their initials or made the tiling patterns for garden areas (see fig. 1.2), they focused on using the tile pieces. It took a while for them to build habits of mind associated with pattern recognition—for example, asking a question like, What stays the same and what is changing?

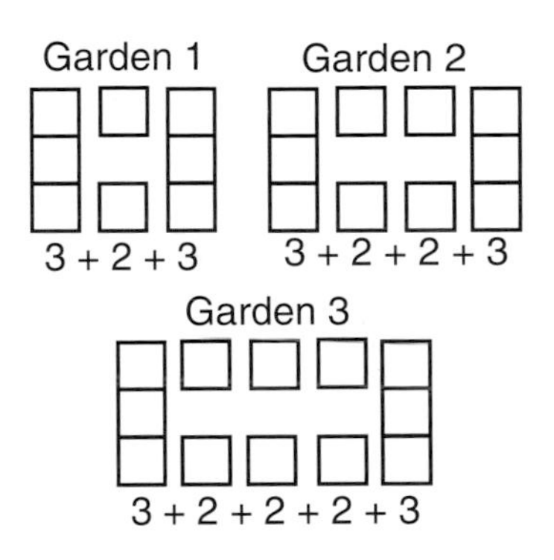

The number of tiles needed for each garden area results in a sequence that is a linear relationship. The recursive rule "Add 2" reflects the constant rate of change of the function, or its slope when graphed. The sum of the values of the three-unit columns, 6, is the start value and, on a graph, is the y-intercept of the function.

With a calculator, enter the first term, 8, followed by +2 = to generate each number in the sequence.

First through Third Elements	Eleventh through Fourteenth Elements	Twenty-second through Twenty-fifth Elements
8	28	50
8	Ans+2	Ans+2
Ans+2	30	52
10	Ans+2	Ans+2
Ans+2	32	54
12	Ans+2	Ans+2
	34	56

The spreadsheet has been completed with the fill-down command. The formulas for determining the number of tiles for garden area 3 and garden area 4 are shown on the spreadsheet. The numerical results of similar calculations are shown for the remaining gardens.

	A	B
1	Garden Number	Number of Tiles
2	1	8
3	2	10
4	3	B3 + 2
5	4	B4 + 2
6	5	16
7	6	18
8	7	20
9	8	22
10	9	24
11	10	26
12	11	28
13	12	30
14	13	32
15	14	34
16	15	36
17	16	38
18	17	40
19	18	42
20	19	44
21	20	46
22	21	48
23	22	50
24	23	52
25	24	54
26	25	56

Fig. **1.3.**

Using recursion on a calculator and on a spreadsheet to find the number of tiles in garden area 25

which also is equivalent to $2n + 6$. Students may find several different statements for the rule, thus encouraging their consideration of equivalent algebraic expressions.

It takes time and many experiences for students to develop their abilities to work expertly with patterns, variables, and functions. The results from the Third International Mathematics and Science Study (TIMSS International Study Center 1997) reveal that the choice and design of tasks, as well as teachers' understanding of the ways students solve problems, are central to successful mathematics teaching (Friel 1998). Rather than suggest a sequence of instructional activities that can be used (see fig. 1.2 for possible problems), we look more carefully here at one problem. We want to consider some of the ways students' written work can enhance our understanding of their thinking about patterns, variables, and functions.

Building with Toothpicks

Goals

- Explore pattern development
- Use a table to organize information
- Generalize a rule that describes how to find the perimeter of the *n*th shape

Materials and Equipment

- A copy of the blackline master "Building with Toothpicks" for each student
- Enough toothpicks for each student to make each of the first four shapes in the pattern shown in figure 1.4

p. 75

Activity

The students may work as partners. The shapes shown in figure 1.4 are made with toothpicks. The students should look for patterns in the number of toothpicks used to make the perimeter of each shape. Present them with the following tasks:

1. Use a pattern from the sequence of shapes to determine the perimeter of the fifth shape in the sequence. Show or explain how you arrived at your answer.
2. Write a formula that you could use to find the perimeter of any shape n. Explain how you found your formula.

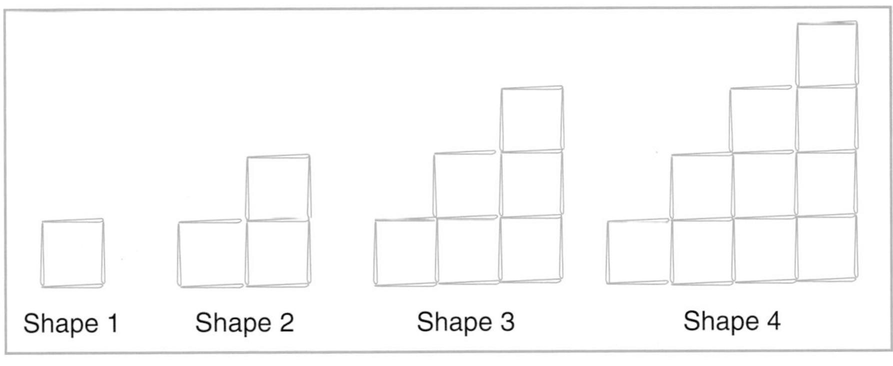

Fig. **1.4.**

A growing pattern made with toothpicks

Discussion

The process of generalizing involves identifying patterns in shapes 1–4, making predictions for a new case (e.g., shape 5, not shown), and eventually describing a general rule for predicting that is expressed in symbolic notation. In figure 1.5 are three examples of student responses to questions that pertain to linear relationships.

Students 1 and 2 made use of the structure of the shapes as a way to help them generalize a description of the pattern. Both students observed that the lengths of two of the sides of a shape had the same value as the number of the shape, so shape 5 would have a length of 5

This activity has been adapted from Burkhardt et al. (2000, pp 1–3).

Building with Toothpicks

The shapes shown below are made with toothpicks. Look for patterns in the number of toothpicks in the perimeter of each shape.

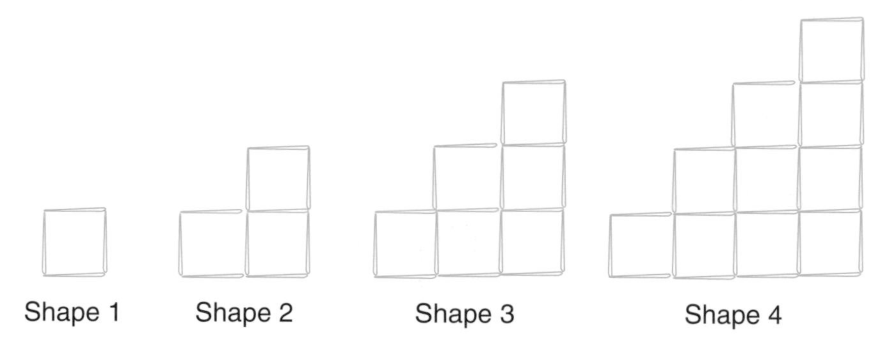

1. Use a pattern from the shapes above to determine the perimeter of the fifth shape in the sequence. Show or explain how you arrived at your answer.
2. Write a formula that you could use to find the perimeter of any figure *n*. Explain how you found your formula.

Student 1: The Use of the Structure of the Shapes

1. Use a pattern from the shapes above to determine the perimeter of shape 5. Show or explain how you arrived at your answer.

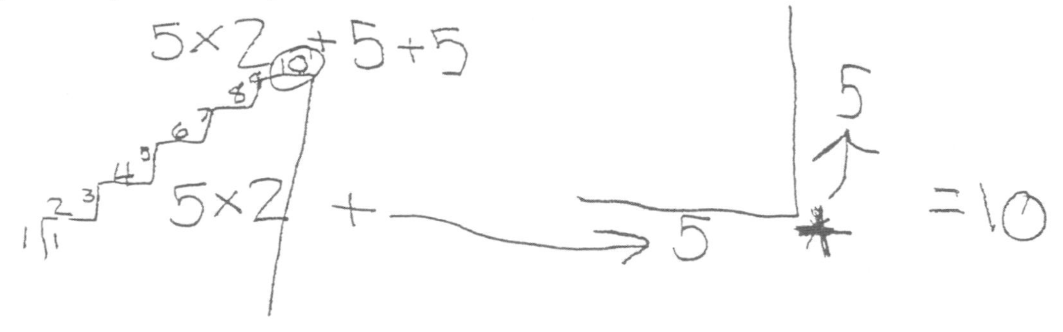

2. Write a formula that you could use to find the perimeter of any shape *n*. Explain how you found your formula.

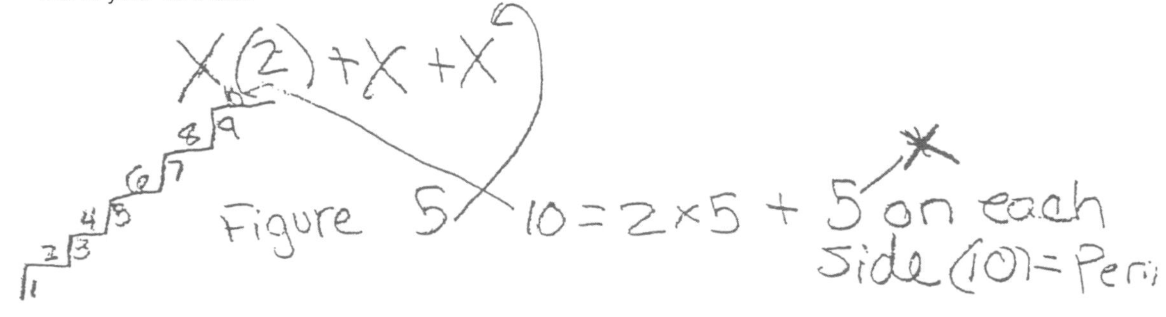

Fig. **1.5.**

Students' solution strategies for the Toothpick Problem

Student 2: The Use of the Structure of the Shapes

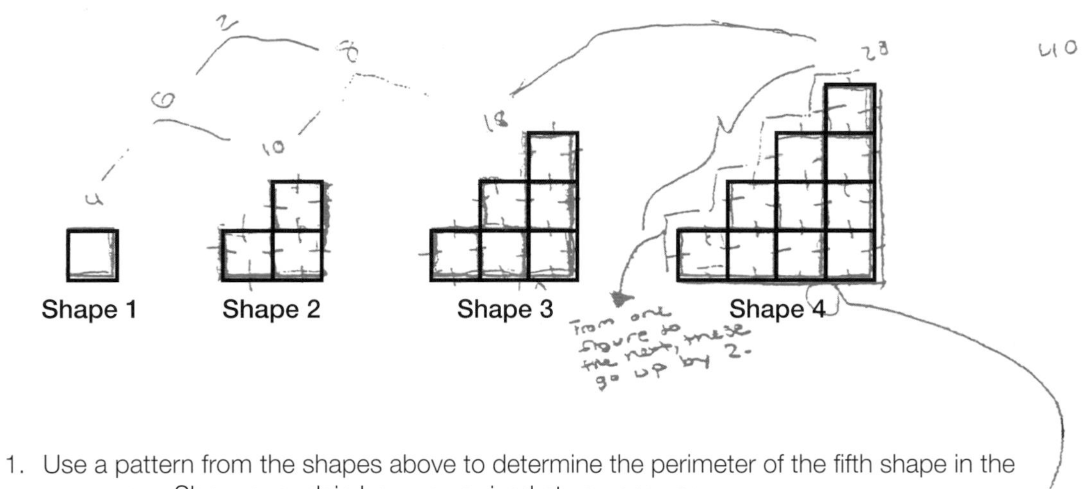

1. Use a pattern from the shapes above to determine the perimeter of the fifth shape in the sequence. Show or explain how you arrived at you answer.

The perimeter is 20.

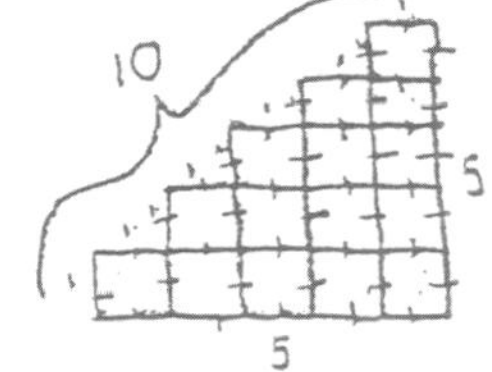

5+5+10=20

The side lengths go up by 1 each time.

2. Write a formula that you could use to find the perimeter of any shape *n*. Explain how you found your formula.

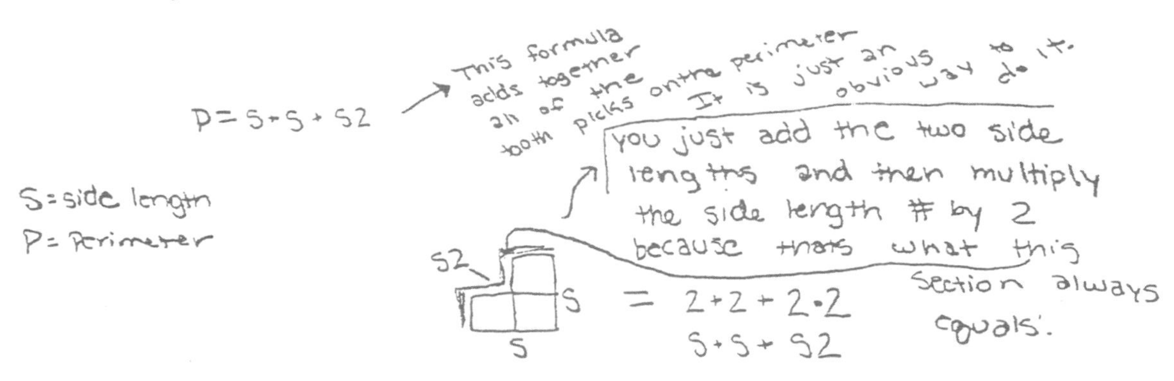

(continued)

Student 3: The Use of Recursion as Part of the Process

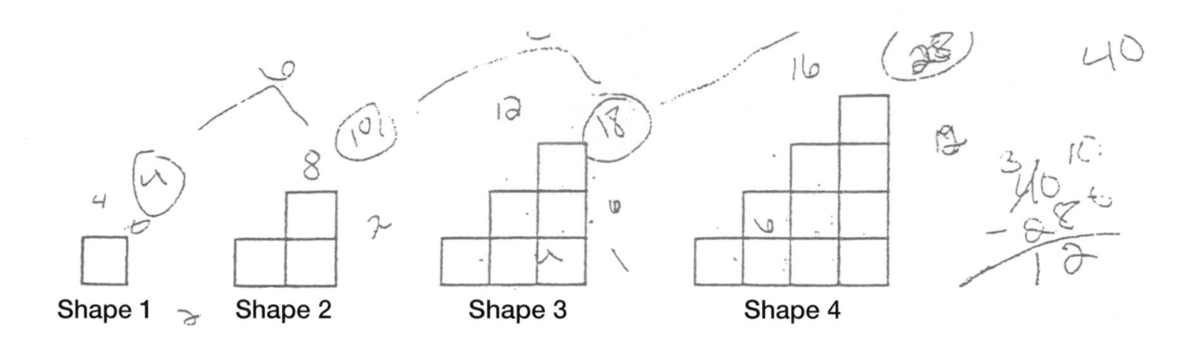

1. Use a pattern from the shapes above to determine the perimeter of the fifth shape in the sequence. Show or explain how you arrived at you answer.

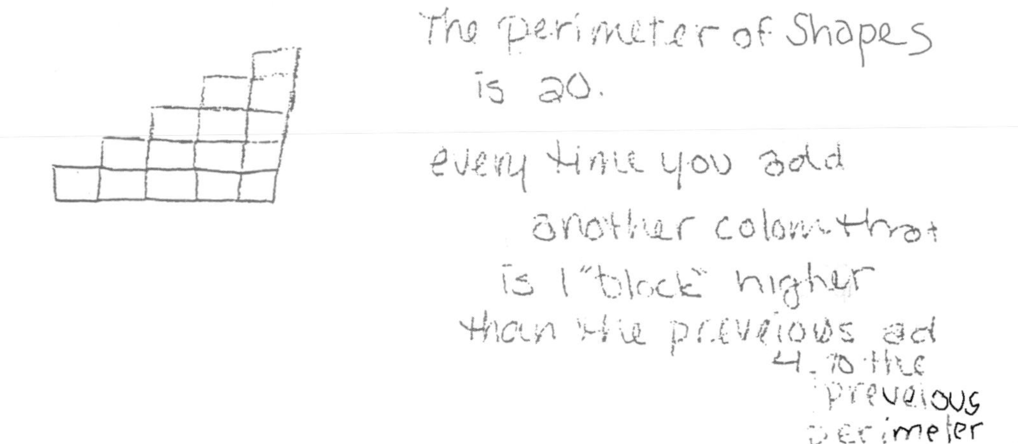

2. Write a formula that you could use to find the perimeter of any shape *n*. Explain how you found your formula.

$y = 4x + 4$ I got this formula because it takes 4 toothpicks to create 1 square. then you add 4 to your perimeter to get the next perimeter.

Fig. **1.5.** (continued)

Students' solution strategies for the Toothpick Problem

units on each of two sides. The length of the third side, because of its stair-step structure, is double the length of either of the other sides, or 10 units for shape 5. Both students used this relationship to express a rule for determining the perimeter of any shape.

Student 3 appears to have used recursion as part of the solution process. She noted that "every time you add another column that is 1 'block' higher than the previous, add 4 to the previous perimeter." In describing a rule, student 3 attempted to express the recursive process she used—adding 4 to the previous element. She noted that there were originally four toothpicks, and she added 4 to the original perimeter to get the next perimeter. Although her formula did not give her the results she wanted, it seems that she was attempting to express the recursive relationship she had been using.

None of the students actually attempted to justify the recursive pattern by exploring what changes and what stays the same in each of the four shapes. Instead of adding the next column of toothpicks, students can be encouraged to make only the perimeter of each shape with toothpicks. The idea is to observe how to rearrange the toothpicks and then add 4 to get the new shape, as shown in figure 1.6.

Notice, also, that the number of toothpicks needed to make the perimeter of each shape results in a sequence that is a linear relationship. The recursive rule "Add 4" reflects the constant rate of change of the function, or its slope when graphed. There is no constant, so on a graph, the y-intercept of the function will be (0, 0).

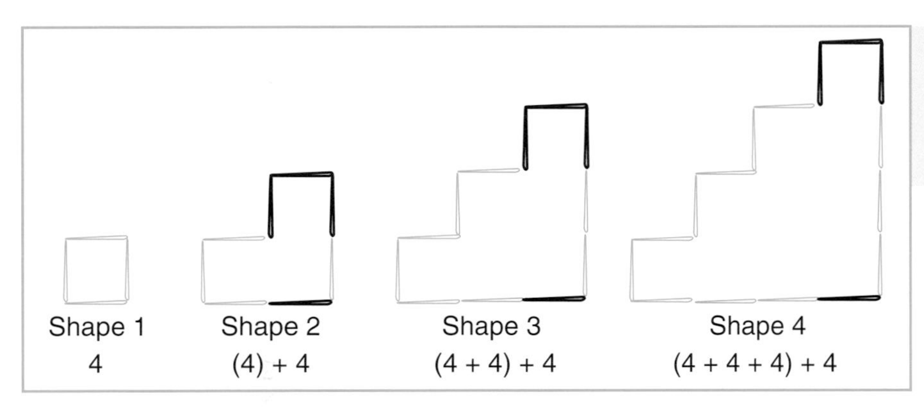

Fig. **1.6.**

Parts of the pattern appear in black to show where the change "Add 4" occurs in each shape.

If this rearrangement is done systematically, students will see that four toothpicks are added the same way each time. This justification eventually permits us to claim that the perimeter is $4n$.

Growing patterns that involve geometric models provide a variety of contexts in which students can explore reasoning using variables and the relationships among two or more variables. It is helpful to encourage students to connect their explanations to the geometric models. If students choose a recursive strategy, they can be asked to justify how the rate of change they have identified occurs as they move from one shape to the next. In choosing an explicit strategy, students 1 and 2 reflected an awareness of how the formula emerges from the structure of the shapes themselves. In the middle grades, the use of techniques such as finding common differences and applying procedures by rote to determine function rules on the basis of common differences is not appropriate unless students can justify their reasoning through evidence from, and reference to, the models of the pattern being considered.

A middle-grades extension of the function-machine model, used in the elementary grades, is function games. Often, using a table for inputs and outputs as part of the game leads nicely to helping students think about using tables as a tool for working with growing patterns. (See Rubenstein [1996], "The Function Game.")

GRADES 6–8

NAVIGATING *through* ALGEBRA

Chapter 2
Analyzing Change in Various Contexts

Discuss what is changing and how it is changing in the following situations. Describe what the change depends on.

1. You are riding your bike for 30 minutes and cover 3 miles; you keep riding your bike for another 30 minutes and cover a total of 6 miles.
2. The volume of the sound you hear from the TV as you move away from it.
3. The temperature of a cup of hot coffee that is left for two hours on the kitchen table.
4. The speed of a car approaching a red light.

Important Mathematical Ideas

Thinking about change is not simple. It requires thinking about several things at once: What is changing? Over what time period is the change occurring? What is the rate at which the change is occurring? Is the change following a pattern? Middle-grades students need to learn to think about these questions mathematically. They need to encounter situations that involve change in order to develop a language for describing and talking about change and to develop representations for the types of change that they may encounter.

In a mathematical context, a variable is a quantity that changes. Quantifiable characteristics that change (vary) might include distance traveled, the height of an object from the ground, elapsed time, and speed. Some important relationships may remain constant (e.g., 60 minutes in an hour). Middle-grades students need opportunities to detect both constant and changing relationships in problem situations by asking both What is changing? and What is not changing? Several important ideas need to be addressed when students explore and analyze change situations (Lamon 1999):

- Change in one variable may or may not be related to change in a second variable (the *dependence* or *independence* of one variable in relation to another).
- Situations may demonstrate constant or varying rates of change. The *direction* in which change is occurring may be *increasing*, *decreasing*, or, in some instances, both increasing and decreasing.

- How quickly one quantity is changing in relation to another (the *rate of change*) also is important to note.
- There are several ways to represent change that involves relationships among variables, including the use of words, tables, graphs, and symbols. It is important to be able to relate and compare these different forms of *representation*.
- A *sketch or graph* shows the relationship of one quantity to another. The shape of the graph provides insights into the nature of the change (e.g., it may be regular, thus reflecting a pattern, or irregular).

What Might Students Already Know about These Ideas?

What do your students know about making and reading tables and graphs? Does the idea that one variable can depend on another have meaning for them?

A variety of activities can stimulate students' interest in studying change and, at the same time, offer insights into students' prior knowledge. Bouncing Tennis Balls, for instance, takes advantage of middle-grades students' physical energy while both assessing their prior mathematical knowledge and introducing them to representing and analyzing change.

Bouncing Tennis Balls

Goal

To assess students'—

- ability to collect data and record data in a table;
- abilty to make a graph to display data using correct labels, scale, and so on;
- recognition of what varies in an experiment;
- ability to name the independent and dependent variables in a problem.

Materials and Equipment

- A copy of the blackline master "Bouncing Tennis Balls" for each student
- Tennis balls, one for each team of four students
- Access to a clock or watch with a second hand
- Centimeter graph paper, a spreadsheet program, or a graphing calculator

p. 76

During the preassessment, one teacher observed her students. She noted which students made graphs correctly, paying attention to how they used the idea of scale to set up the time and distance axes. She also listened to students' conversations about their graphs, attending to comments that indicated the students realized that the number of bounces depends on the length of time the ball is bounced and that patterns develop when the ball is bounced in a consistent way.

Activity

In teams of four, students bounce a ball to solve this problem:

> How many times can each team member bounce and catch a tennis ball in two minutes?

A bounce is defined as dropping the ball from the student's waist. One student keeps the time while the second student bounces and catches the ball, the third student counts the bounces, and the fourth student records the data in a table showing both the number of bounces during each ten-second interval and the cumulative number of bounces. Each trial consists of a two-minute experiment, with the number of bounces recorded after every ten seconds (or twenty seconds for fewer data points). The timekeeper calls out the time at ten-second intervals. When the time is called, the counter calls out the number of bounces that occurred during that ten-second interval. The recorder records this count and keeps track of the cumulative number of bounces.

The same process is followed by each student, with the students rotating roles, so that each student can collect a set of data. All the students must bounce the ball on the same surface (e.g., tile, carpet, concrete) because differences in the surface could affect the number of bounces.

Once the data have been collected, each student prepares a graph showing the cumulative bounces over two minutes. This graph can be constructed by hand, by using a graphing calculator, or by using a spreadsheet, depending on the students' experiences and on what information the teacher wants to gather about what the students know and are able to do.

This activity has been adapted from Jones and Day (1998, pp. 18–19).

Discussion

The data from one student's experiment are recorded in table 2.1. The graph in figure 2.1 was made using the graphing calculator, and the graph in figure 2.2 was made using a spreadsheet for the sample data set.

Table 2.1
A Sample Data Set for Bouncing Tennis Balls

Time (Seconds)	Number of Bounces during Interval	Cumulative Number of Bounces
0	0	0
10	11	11
20	11	22
30	9	31
40	10	41
50	11	52
60	10	62
70	11	73
80	11	84
90	10	94
100	10	104
110	10	114
120	10	124

Fig. **2.1.**

A graph made using a graphing calculator

The window shows the scale.

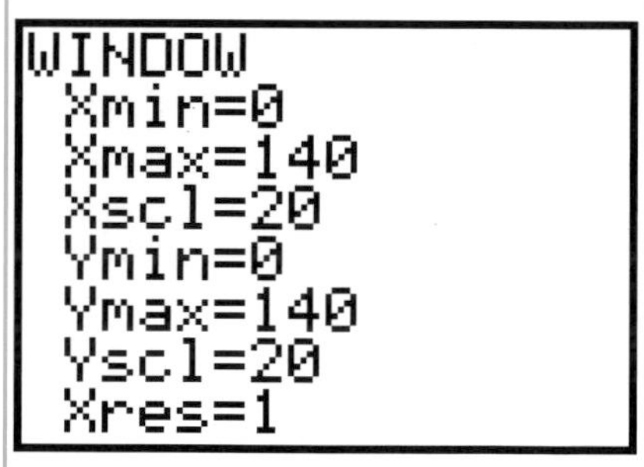

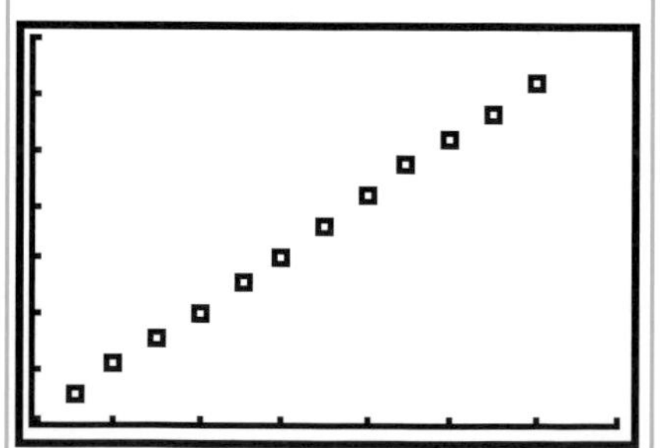

Students present their results to classmates by showing their graphs. The discussion can involve what the students found easy and what they found difficult in completing this task. Students' discussions can be revealing: Can the students identify what varies in the experiment? Do they comment on the dependent and independent variables either implicitly, in their conversations about the graphs, or explicitly, using correct terminology? Do they discuss whether the points should be connected with a line? The numbers of bounces are discrete data, so they should not be connected. Decisions about the scale for each of the axes are important; do the students understand what the graphs would look like if the scales changed? When directed to sketch lines on their graphs in order to notice trends, do they demonstrate some sense that the steepness of a line is related to the number of bounces per second? Your observations related to these and other questions will yield information about what your students appear to know and are able to do that will guide you in making instructional decisions.

An extension of this activity would be for each student to conduct an experiment using, for example, concrete floors and then carpeted floors to investigate the effect of differences in the surfaces.

Selected Instructional Activities

Highlighted here are some fundamental components of a curriculum that addresses content and develops students' understanding by focusing on analyzing change.

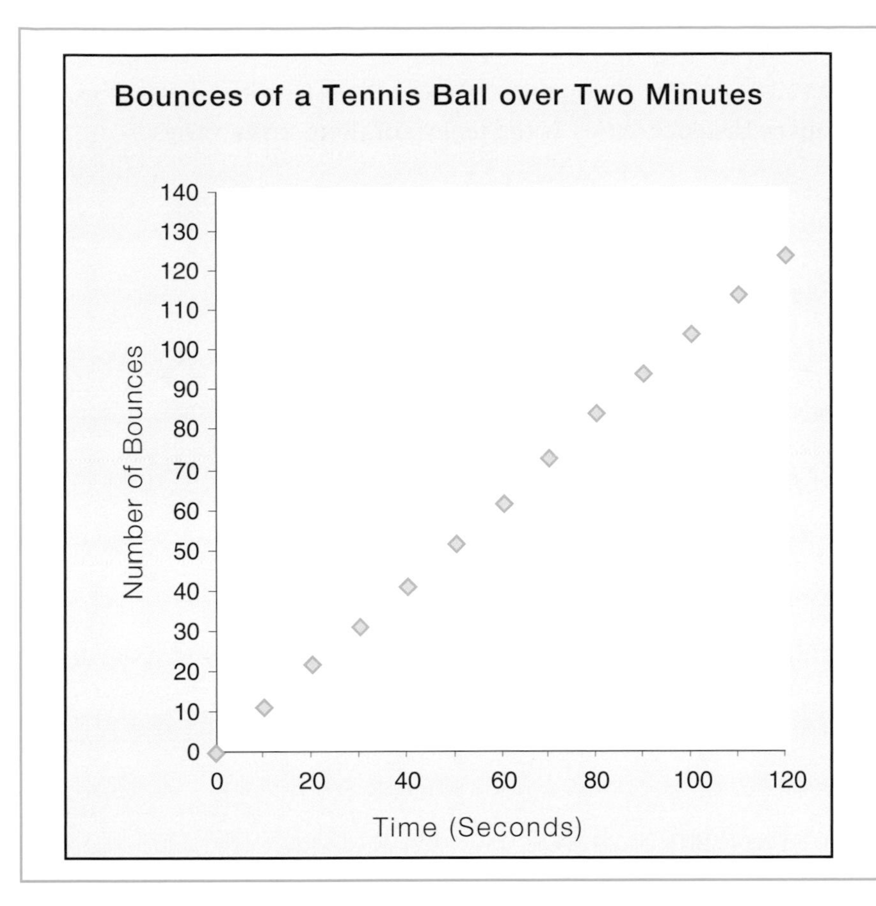

	A	B
1	Time (s)	No. of Bounces
2	0	0
3	10	11
4	20	22
5	30	31
6	40	41
7	50	52
8	60	62
9	70	73
10	80	84
11	90	94
12	100	104
13	110	114
14	120	124

Fig. **2.2.**

A graph constructed using a spreadsheet

Building a Sense of Time and Its Relation to Distance and Speed

Initially students need to become aware of their own understanding of time, change over time, and the use of new kinds of measure (i.e., rates). Posing such questions as those listed below focuses their attention on these ideas (adapted from Kleiman et al. 1998).

- How do you measure time? Distance? Speed?
- Give an example of something that might be able to travel at two feet per second.
- What is the difference between traveling at two feet per second and two feet per minute or two feet per hour?

Students can also explore different activities that test their sense of time. They can do the following activities in pairs; in each instance they may want to observe if they overestimate or underestimate the time and try the task again.

- Clap your hands so you clap exactly one clap per second for ten seconds.
- Turn a page in a book at exactly one page every two seconds for twenty seconds.
- Sit still for thirty seconds, letting the timer know by raising your hand when you think thirty seconds has passed.
- Walk at the speed of one foot per second for fifteen seconds.
- Walk the length of your classroom in exactly ten seconds. At what speed were you traveling?

In this context, distance is how far the object or person moves (travels). Speed is how fast the object or person is moving (traveling). Both are described in terms of direction. Distance is measured in such units as feet, miles, or kilometers. Speed is measured in relation to time using units such as meters per second or miles per hour.

One teacher reported that many students initially misjudged time. Before she let them explore the activities in teams, she had them sit for an undisclosed time (e.g., 30 seconds) and make guesses about the amount of time that had elapsed.

Connecting Graphs to Stories

In "Walking Strides," students examine how the time required to walk a given distance varies as the length of their stride varies.

Walking Strides

Goals

- Recognize what varies in the experiment
- Identify the independent and dependent variables
- Use a table to organize information
- Make a graph to display data using correct labels and scales
- Consider how the steepness (slope) of a line relates to distance covered in a given time

Materials and Equipment

- A copy of the blackline master "Walking Strides" for each student
- An established distance, with one-quarter of the distance marked off
- Access to a clock or watch with a second hand or to a stopwatch
- Centimeter graph paper or a graphing calculator

p. 77

Activity

Working in teams, students set up an experiment in which one student, the walker, walks one-quarter of a given distance (e.g., one-quarter of 80 or 100 feet). The walker chooses to use a short, regular, or long stride and uses this stride for the marked-off quarter of the predetermined distance. (We are assuming that short, regular, and long strides result in different paces.) A second student, the timekeeper, records the time it takes the walker to walk this distance. The walker repeats this procedure for each of the three strides. The walker and the timekeeper then switch places and repeat the experiment.

Once a team has completed gathering the data, each student uses this information to estimate the time it would take to walk half the distance, three-quarters of the distance, and, finally, the total distance, using each of the three strides and assuming that each student's pace is constant for a given stride. These estimates are also recorded. The next step is to make graphs of these data, plotting the points by hand on a coordinate grid or using a graphing calculator.

Possible definitions of strides:

- A short stride: the walker walks toe to heel with no space between the feet.
- A regular stride: the walker leaves the length of about one of his or her feet between the toe of one foot and the heel of the other foot.
- A long stride: the walker leaves the length of about two of his or her feet between the toe of one foot and the heel of the other foot.

Discussion

One student's data recording is shown in figure 2.3. The next step is to make graphs of these data, plotting the points by hand on centimeter graph paper or using a graphing calculator (see fig. 2.4). Students should consider what features are the same and what features vary for the graphs.

Walking Strides
(80 feet total distance)

Distance	Time in Seconds for Different Strides		
	Short	Regular	Long
1/4 distance (20 feet)	20	9	6
1/2 distance (40 feet)	40	18	12
3/4 distance (60 feet)	60	27	18

Fig. **2.3.**

One student's recording of her data for Walking Strides

Fig. **2.4.**

The time and distance for three different strides graphed on a graphing calculator

L1 is the distance in feet: L2 is the time with a short stride; L3 is the time with a regular stride:

L1	L2	L3 1
0	0	0
20	20	9
40	40	18
60	60	27
80	80	36
------	------	------

L1(6)=

The window shows the scale:

```
WINDOW
 Xmin=0
 Xmax=100
 Xscl=10
 Ymin=0
 Ymax=80
 Yscl=10
 Xres=1
```

Three graphs are shown below on one grid. Pressing TRACE highlights the distance, 40 feet, and the time, 18 seconds, to travel this distance with a regular stride.

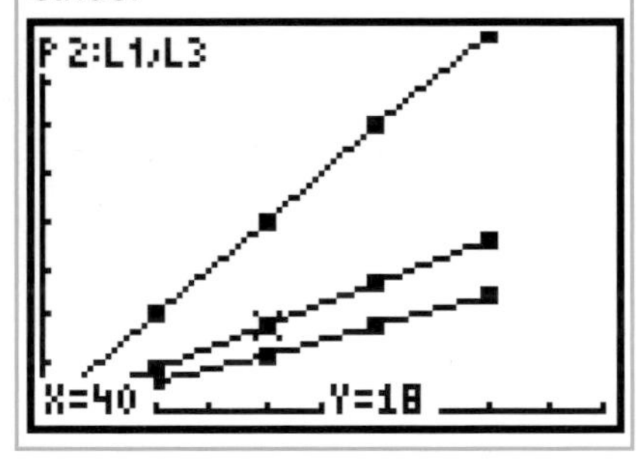

The concepts of independent and dependent variables and rate of change can be revisited. In addition, students can spend some time discussing the "steepness," or the slope, of the different lines and what the slope means in terms of distance traveled. Ask questions like the following: What does the difference between the steepness of two lines tell us? What could we say about distance traveled if a line were vertical? If a line were horizontal? Overall, from the appearance of the graphs, what stories can we tell about walking using strides of different sizes?

Students could also consider stories that involve walking or other contexts to explore distance-time relationships or other situations in which time is the independent variable. In the activity From Stories to Graphs, students sketch graphs to portray such situations.

From Stories to Graphs

Goals

- Identify the independent and dependent variables in problems
- Sketch a graph to represent a story context that involves change over time

Materials and Equipment

- A copy of the blackline master "From Stories to Graphs" for each student

p. 78

Activity

Distribute copies of the blackline master, and have the students complete the following tasks presented on it:

1. In a walking experiment, Josephine walked a total distance of 40 feet. At the halfway point, she had walked for 25 seconds. She stopped for 5 seconds to tie her shoe and then continued walking for 25 more seconds. Sketch a graph that shows Josephine's distance from the starting point over time. (See fig. 2.5.)

2. You are gathering data in the school cafeteria from 8:00 A.M. to 3:00 P.M. Sketch a graph that tells a story about the number of cans of soda in a vending machine over that time. Write a paragraph that tells the same story in words. (See fig. 2.6.)

Fig. **2.5.**

A graph of Josephine's walk

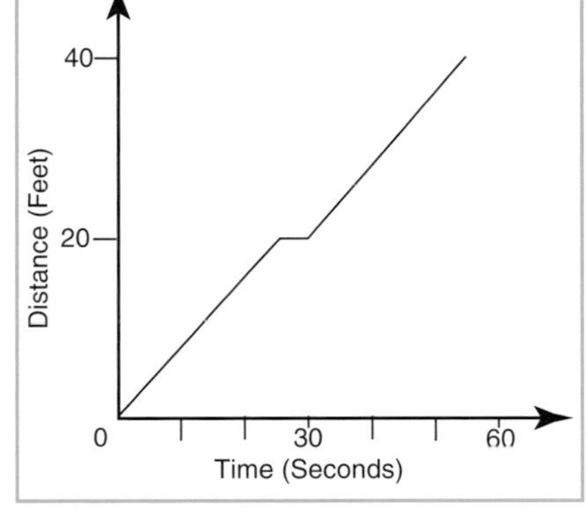

There isn't much action right away, but then a few teachers come in to get sodas. Things continue to be slow; at various times, a student or two and a few teachers stop by for sodas. Then around 10 A.M. during a small rush between classes, several students get sodas. Just before noon, there is a steady stream of students and teachers visiting the soda machine. After lunch break is over, there is another lull. Then, at a class bell, there is another small rush for sodas right before the last class. Students are dismissed, and several stop to get sodas on their way out.

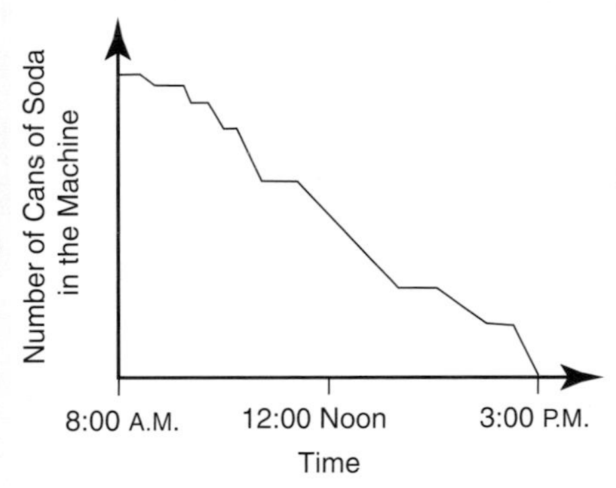

Fig. **2.6.**

A graph of the number of cans of soda in a vending machine and an explanatory story

3. You are mowing the lawn. As you mow, the amount of grass to be cut decreases. You mow at the same rate until about half the grass has been cut. Then you take a break for a while. Then, mowing at the same rate as before, you finish cutting the grass. Sketch a graph that shows how much uncut grass is left as you mow, take your break, and finish mowing. (See fig. 2.7.)

Fig. **2.7.**

A graph representing the amount of grass left to be cut as a lawn is mowed

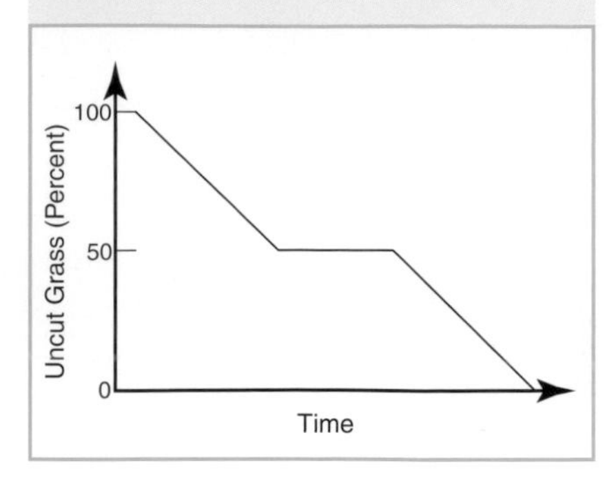

Discussion

You may find, when it comes to drawing graphs, that the students are challenged by the request to "sketch a graph" when just a few or no numbers are specified. Students may want to plot points and connect them to make a graph and may be confused without a numerical label on each axis.

Telling Stories from Graphs

As students gain a greater understanding of analyzing change by making graphs, they can be asked to make up stories for graphs they are given or to read graphs to answer questions about a story "in progress," as in the activity From Graphs to Stories.

From Graphs to Stories

Goals

- Identify the independent and dependent variables in problems
- Describe the story context on the basis of a graph that displays changes over time

Materials and Equipment

- A copy of the blackline master "From Graphs to Stories" for each student

Activity

Distribute copies of the blackline master, and have the students complete the following tasks presented on it:

p. 79

1. John and his father participate in a 100-meter race. John started the race 3 seconds after his father began to run. The graph in figure 2.8 provides information about how far John and his father ran over time. Write a story about who won the race; be descriptive about how the race was run. If the two lines describing how each person ran were parallel, what would the graph tell you about who won the race? (See figure 2.8)

Both John and his father ran a steady pace for the duration of the race. John's father finished the race at about 17 seconds; John finished before his father, at about 16 seconds. If the two lines were parallel, it would mean that both John and his father had run the race at the same pace and completed it in the same amount of time. However, John's father would have won because he started before John.

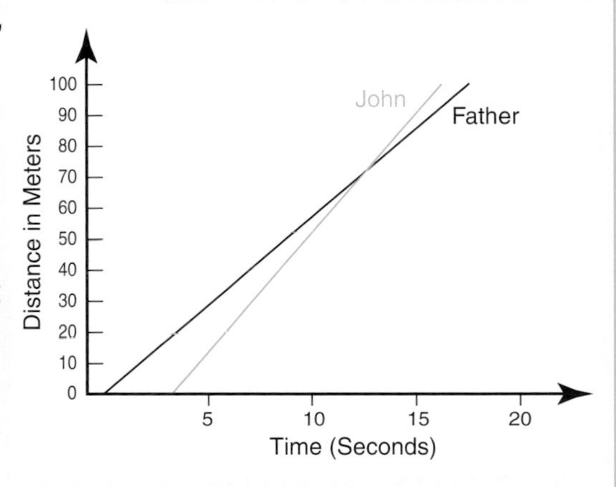

Fig. **2.8.**

A graph and story for a 100-meter race

For more ideas, see Nickerson, Nydam, and Bowers (2000) or Lambdin, Lynch, and McDaniel (2000).

2. The graph in figure 2.9 represents the relationship between the profit and the amount of lemonade sold at a lemonade stand. Write a story about how the lemonade stand's profit is determined. Include an explanation of what is indicated when the line is below zero and when the line crosses the horizontal axis. (This graph assumes that the seller is not paid and that there is no overhead.)
3. The graph in figure 2.10 represents a flag being raised on a flagpole. Write a story that describes what is happening to the flag, gives an estimate of the height of the flagpole, and explains the shape of the graph.

Fig. **2.9.**

A graph and story for sales and profits for a lemonade stand

The profit starts out below zero before there is any lemonade sold. This is because you need cups, lemons, sugar, ice cubes, and water to make the lemonade. Once glasses of lemonade are sold, there is a point when you have sold enough lemonade to pay for your costs (when the line crosses the horizontal axis). Then you start to make a profit. The profit decreases suddenly because you have to buy more supplies. Then you continue to make a profit at the same rate as before.

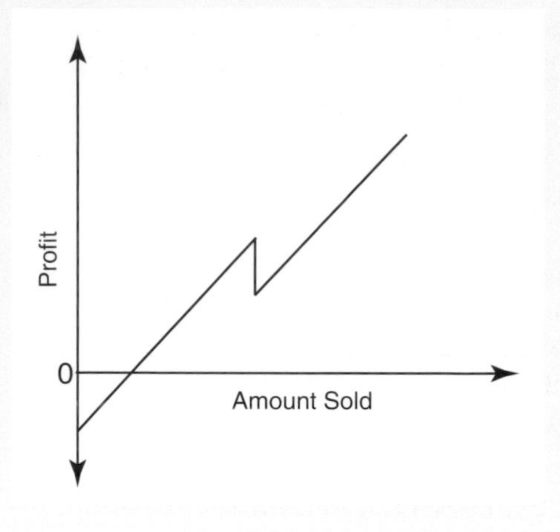

Fig. **2.10.**

A graph and story representing the raising of a flag

When you raise a flag, you pull steadily with both hands together for a short time, then move your hands up the rope that is used to raise the flag. The graph shows a steady pace at raising the flag in a cycle of pull–move hands–pull–move hands until the flag reaches the top of the flapole. if we estimate that the flag is raised about three feet at each stage, the flagpole is about fifteen feet in height.

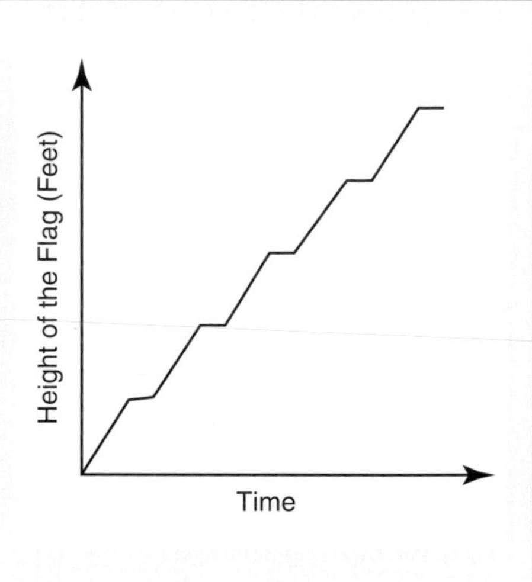

4. Graphs can be used to depict the story of a race. Figure 2.11 includes a graph that shows a swimming race that occurred between two middle-grades students. Write a story that describes what happened in this race.

Fig. **2.11.**

A graph and story representing a swimming race

Cindy started off swimming faster than Kelly; she was ahead of Kelly when they completed the first lap across the pool. On the way back, in her second lap, Cindy slowed down and completed the race in more than 60 seconds. Kelly seemed to spurt ahead, moving past Cindy near the end of the race, and finishing the race in under 60 seconds.

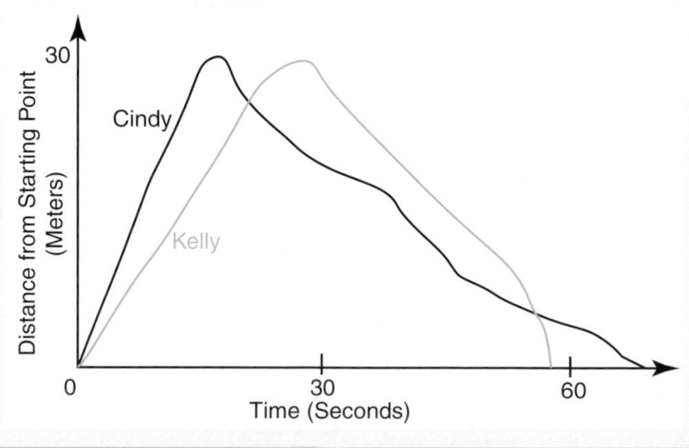

Discussion

You may find that your students are successful at interpreting the graphs they are given, even when the graphs have no numerical scales on the axes. They may understand what each axis represents and how to interpret the changes each line, or curve, represents.

Puzzling about Graphs of Special Kinds of Situations

Not all graphs make sense as real-life situations. What does it mean for a graph to be impossible? Generally, it means that the graph contradicts our understanding of mathematics or science. For example, speed doesn't change instantly; if a car traveled at fifteen miles per hour and then at twenty-five miles per hour, at some point it had to have been traveling twenty miles per hour. Figure 2.12 is an example of an impossible graph relating distance and time in which gaps occur; in the real world, this graph would have to be continuous rather than discontinuous.

Figure 2.13 shows two graphs for students to consider. Ask the students to make up stories for each of these graphs. If they conclude that either graph is impossible to relate to a real-world situation, have them explain why and then make changes so that it portrays a possible situation about which they can tell a story.

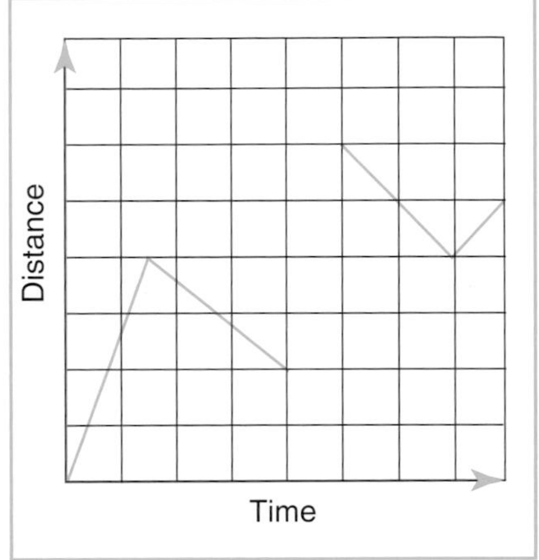

Fig. **2.12.**

An example of an "impossible" graph

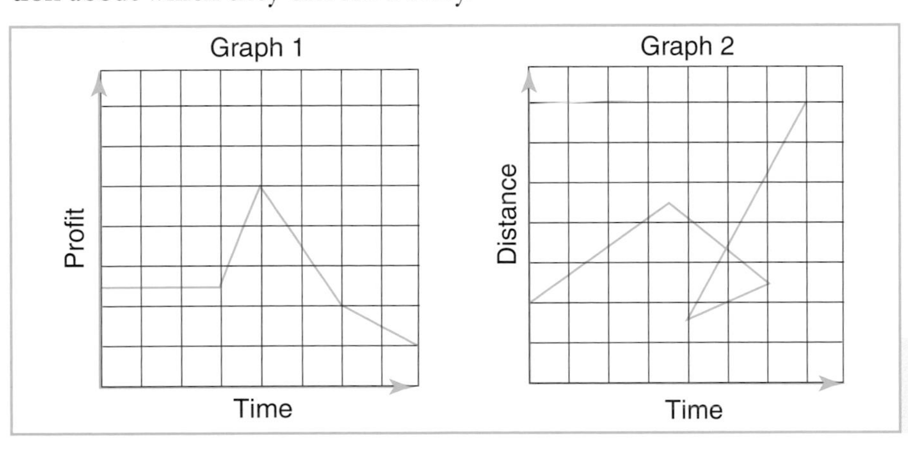

Fig. **2.13.**

Two graphs for students to interpret

Using a motion detector or a motion ranger connected to a graphing calculator, students can focus on change in the appearance of speed represented on a graph. They can create different linear graphs that reflect their movements at different constant speeds. In doing so, students not only gain an understanding of graphical representations of speed, they also develop intuitive ideas about slope.

Graphs of distance versus time and speed versus time pose different kinds of challenges. It is important to help students distinguish between the two situations. Most students confuse distance with speed. Students need opportunities to make and to read both kinds of graphs in which speed remains constant for different intervals of time. Although students can make sense of both kinds of graphs (speed over time and distance over time), it is important to be aware of the complexities involved in doing so.

Consider the task Tony's Walk (Lane 1993; Silver and Lane 1993) shown in figure 2.14. Students are asked to write a story about Tony's walk. In their stories, they describe what Tony might have been doing at the different times. Figure 2.15 shows the responses of three different

See Beckmann and Rozanski (1999), Johnson (2000), and Lapp and Cyrus (2000), for ideas about graphing.

Students 2 and 3 marked off intervals on the speed axis of the graph to show 1, 2, 3, 4, 5, 6, and 7 MPH.

students. Student 1 estimated walking speeds and told a story that provided reasons for the changes in the Tony's speed. Student 2 responded by writing a log of times and what was happening with Tony at those times. Notice that this student seems to be aware of the transition time needed by Tony to reach his walking pace; she includes times such as 12:03 or 1:01 in her log. The response by student 3, although briefer than the responses of the other two students, demonstrates an understanding of the context and what is happening to Tony's speed over the three hours of this walk.

Fig. **2.14.**

The task Tony's Walk, for which students write stories

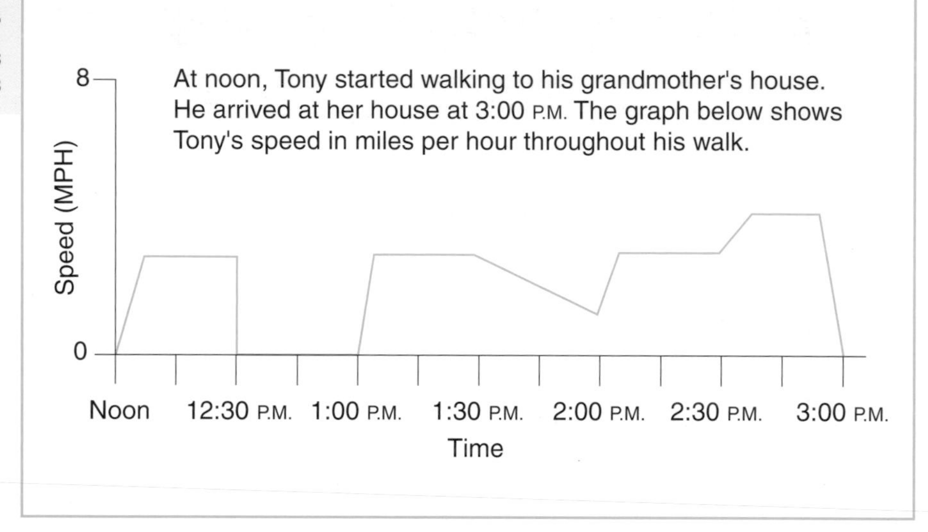

Fig. **2.15.**

Three students' stories about Tony's Walk

Student 1

Tony left his house at noon. From noon to 12:30 he was walking about 3½ miles per hour because he was trying to catch ants. At about 12:30 Tony stopped to rest and eat lunch. Then at 1:00 he started walking again, he was walking about 3½ miles per hour until about 1:30 because he was daydreaming. At 1:30 he saw five birds hopping along the sidewalk so he watched them while he was walking which slowed him down a little. Then at 2:00 he finally caught up with his regular pace until 2:30 when he realized the time then he ran the rest of the way.

Student 2

At 12:00 NOON he left home

At 12:03 he slowed his pase down

At 12:25 he decided to do something with friends until 1:00

At 1:01 he Kept his speed at a normal rate

at 1:30 he got tired slowed down

At 2:00 he looked at his watch and relized the time

So he sped up

At 2:30 he realized that he had Only 30 minutes

to get to granny's

At 2:35 he jogged to granny's. 3:00

Student 3

One day tony decided he wanted to go to his grandmother's he started walking at noon til 12:30 he rested and started agin at 1:00 Then he started getting tired at 1:30 and slowed down a little bit. At 2:00 he speeded up at 2:30 he went even faster at three he was where he wanted to go.

Linking Verbal Descriptions, Tables, and Graphs

In the study of algebra, it is important to connect tables with graphs and symbols and to consider how these tools help students understand functional relationships in context. Problems like the ones that follow help students focus on these tools, their relationships, and what information each provides.

1. Camping is popular with many people. When campers go to a campsite, they pay a fee for using the site and facilities. A campground keeps a record of the fees it charges in order to determine income. Here is a table of data for a campground where the fee for a campsite is $14.50:

Number of campsites	1	2	3	4	5	6
Total campground fee	$14.50	$29.00	$43.50	$58.00	$72.50	$87.00

(*a*) Make a graph of these data. (See fig. 2.16.)

Fig. **2.16.**

A graph of the campsite-fee data

L1 is the number of campsites; L2 is the total fees collected for that number of campsites.

L1	L2	L3 2
1	14.5	------
2	29	
3	43.5	
4	58	
5	72.5	
6	87	

L2(7) =

The window shows the scale:

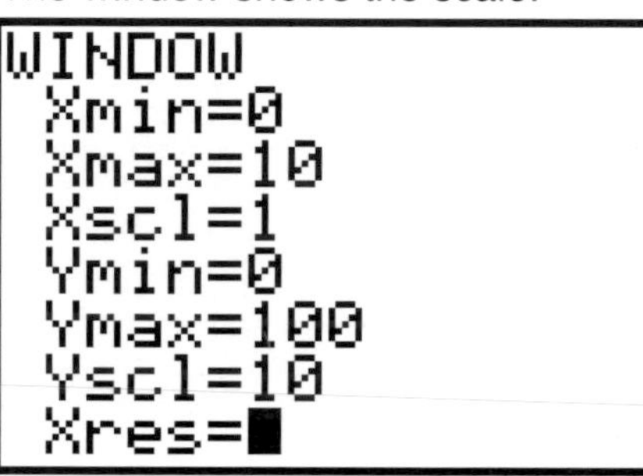

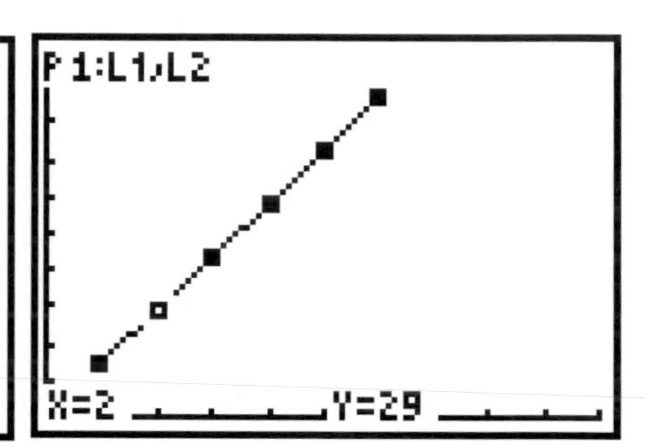

(*b*) Would it make sense to connect the points on the graph? (no) Explain your reasoning. (It would not make sense to connect the points on the graph, since the fees are for entire campsites; no partial fees are paid because no partial campsites can be rented. When we graph distance traveled [on the vertical axis] versus time [on the horizontal axis], we can find the distance traveled for all values of time, not just for whole numbers. The graph reflects a continuous function. All points are connected [see fig. 2.8]. Graphs like those in figure 2.16 do not contain connected points because they display discrete data. A dashed line may, however, be drawn connecting the points for discrete data in order to use the line to help identify trends or make predictions for other cases not shown by the graph.)

(*c*) Using the table, describe the pattern of change that is occurring. How does this pattern of change relate to your graph? (From the table, we can see a constant pattern of change—"Add $14.50." The graph is linear.)

(*d*) Write a rule that relates the number of campsites and the total campground fees so that you could calculate campground fees for any number of campsites. (*Rule:* Fee = $14.50 × number of campsites.)

2. The Jamestown Hornets, the undefeated basketball team, traveled by bus to their last game. They drove round-trip 420 miles at an average speed of 60 MPH.

(*a*) Make a table and a graph of the time and distance traveled by the team. (See fig. 2.17.)

(*b*) Would it make sense to connect the points on the graph? Explain your reasoning. (It would make sense to connect the points on the graph because distance is a continuous measure.)

L1 is the number of hours; L2 is the number of miles traveled based on an average speed of 60 MPH.

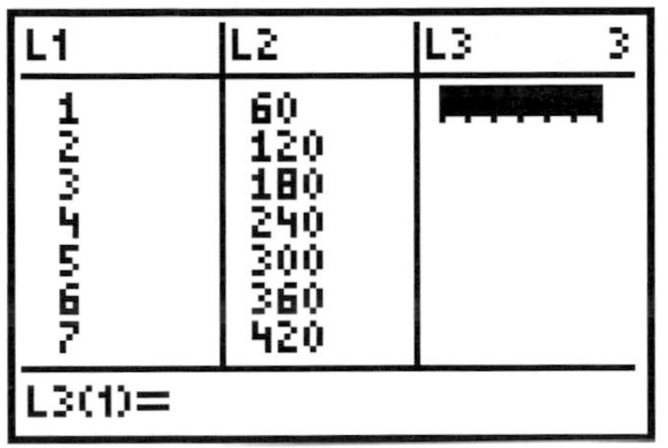

The window shows the scale:

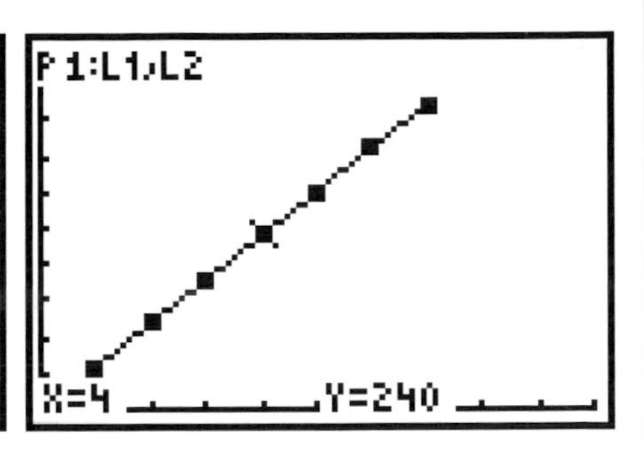

Fig. **2.17.**

A table and graph of the Hornets' travel data

(*c*) Explain how you can estimate the distance traveled in two hours from the table. From the graph.

(*d*) Explain how you can estimate the distance traveled in 3 3/4 hours from the table. From the graph.

(*e*) Write a rule that relates time and distance that you could use to help you calculate the distance traveled, given any time on this trip. (*Rule:* Distance [in miles] = 60 miles per hour × time [in hours].)

Notice that these two problems pose a question about whether or not the points on a graph should be connected with a line in the given situations. Students need opportunities to explore discrete and continuous phenomena; these problems offer such opportunities. In addition, when points can be connected, it is important to ask, "Connected in what way?" For example, consider the ways of connecting two points shown in figure 2.18.

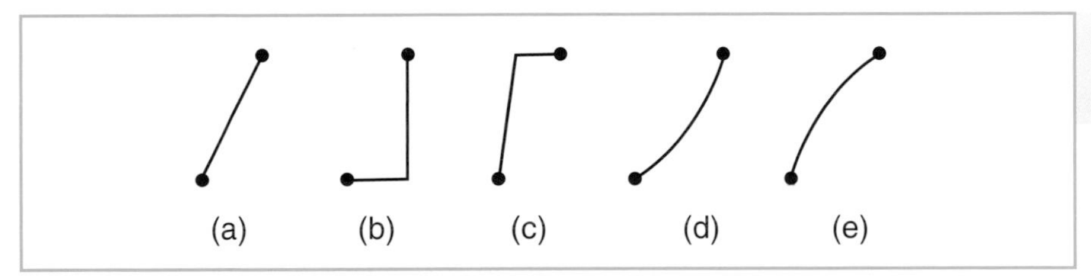

Fig. **2.18.**

Different ways of connecting two points

If we consider the trip that the Jamestown Hornets took on the bus and the graph displaying points for cumulative distance traveled for each hour of the trip, the points may be connected with a straight line. Suppose that between hour 2 and hour 3, the two points were connected as shown in figure 2.18c; what could we tell about how the distance was covered? (The bus traveled the distance in less than one hour and then was stationary for part of the hour.) Is it possible that the two points could be connected as shown in figure 2.18b? (This line suggests that the bus traveled no distance for an hour, then traveled the distance all at once. It is not possible for the two points to be connected as shown.) Is it possible that the two points could be connected as shown in figure 2.18d or e? (Both are possible.) What would each of these lines tell us about the way the distance was covered during this hour of the trip? (Both curves suggest the way the distance was covered. In (*d*), the bus started traveling slowly and continuously accelerated; in (*e*), the bus traveled fast at first and then continuously slowed as it approached

Grades 6–8

Navigating through Algebra

Chapter 3
Exploring Linear Relationships

In various real-world situations, the value of one variable depends in some predictable way on the value of another. To describe this kind of relationship of dependence, we use phrases like "attendance depends on ticket price" or "height is a function of age." In mathematics, the word *function* is used to describe a unique relation among variables in which for each value of the input variable, there is one and only one related value of the second variable, the output variable.

Two of the central purposes of algebra are (1) to develop skill in, and an understanding of, the ways that functions can be used to describe and study relationships among quantitative variables and (2) to develop an understanding of the behaviors of various families of functions. We can use tables of value pairs (input, output), graphs of these data pairs, and symbolic rules to represent functions.

Important Mathematical Ideas

Building the concept of linearity is central to middle-grades mathematics. This concept includes constant unit rates, slope, and *y*-intercept. In a linear relationship, there is a constant rate of change between two variables. The change in the dependent variable that is associated with a given change in the independent variable will stay the same over the range of inputs to the function. For example, every time the diameter of a circle increases by one unit, the circumference increases by π units; this relationship is linear. Graphs of linear functions are straight lines.

Linear relationships form one of the most important and basic families of functions that students study in the middle grades. The ideas that are important for students to consider include the following:

- The relationships among variables can be recognized and represented in a variety of ways, including in words, tables, graphs, and symbols.
- Patterns in which each term is obtained by adding a fixed number to the previous term are examples of linear relationships (see chapter 1 in this book).
- Linearity is associated with a constant rate of change between two variables.
- The function rule for linear functions is $y = mx + b$, where m is the slope, or rate of change between the two variables, and $(0, b)$ is the y-intercept.
- The slope of a line and the y-intercept can be found from a graph, a table, or an equation.
- Given a specific problem described using $y = mx + b$, it is possible to identify how the values of m, x, b, and y relate to the problem.
- Changes made in either the slope or the y-intercept affect the various representations, that is, tables, graphs, and symbolic expressions.
- A solution common to two linear equations can be found by creating tables for or graphing the two functions.

What Might Students Already Know about These Ideas?

As students work with formulas and explore patterns, they are encountering functions. Their initial work in these areas often focuses on linear functions, although the concept may not be formalized. In moving to a formal consideration of linearity, it may be helpful to assess students' understanding of the relationship among tables, graphs, and symbols. One possible way to do so is to have the students complete the following task.

Missing Values

Goal

To assess students'—

- ability to make a graph to display data using correct labels and scales;
- recognition that a constant rate of change is evident in the pattern established by the data;
- recognition of the relationship among a table, a graph, and a symbolic expression.

Materials and Equipment

- A copy of the blackline master "Missing Values" for each student
- Graphing calculators (optional)

p. 80

Activity

A piece of paper with an incomplete table (see fig. 3.1) showing number pairs was left on a student's desk after a mathematics class. Using what you know about patterns, answer these questions:

1. Predict the missing values for y, and record them in the table. Describe how you made your predictions.
2. Graph the data in the table on a coordinate grid.
3. Describe a general rule to help you determine the value for y if you are given the value for x.
4. Use your rule to find the value for y when $x = -1$. Show this point on the line.

Fig. **3.1.**

The incomplete table used in Missing Values

x	y
2	___
3	___
4	16
5	20
6	24
7	___
8	32

Discussion

This activity can reveal students' understanding of the relationship between a table and a graph, their awareness of constant rate of change, and their ability to write a function to describe how y depends on x. How do students set up the coordinate grid? What labeling, if any, do they use? How do they reason about predicting missing values? Do they note the constant rate of change occurring in the table? Do they use the graph to locate points? What is the reasoning behind the rule they derive? Do they seem to focus on a recursive or an explicitly defined functional relationship? Are they able to extend their understanding to include points that are not on the table—in this case, a point below the x-axis?

The graphing calculator screens in figure 3.2 display a solution to this problem.

Selected Instructional Activities

Several essential concepts can be developed as part of an exploration of linearity. Instead of proposing a sequence of activities that support the development of each concept in depth, we have selected activities that highlight the variety of important ideas to be addressed. Although the organization of these activities appears linear, in reality, the concepts are interrelated and are developed dynamically through a variety of experiences.

Fig. **3.2**.

Graphing-calculator screens for the Missing Values problem

L1	L2	L3 3
2	8	
3	12	
4	16	
5	20	
6	24	
7	28	
8	32	

L3(1)=

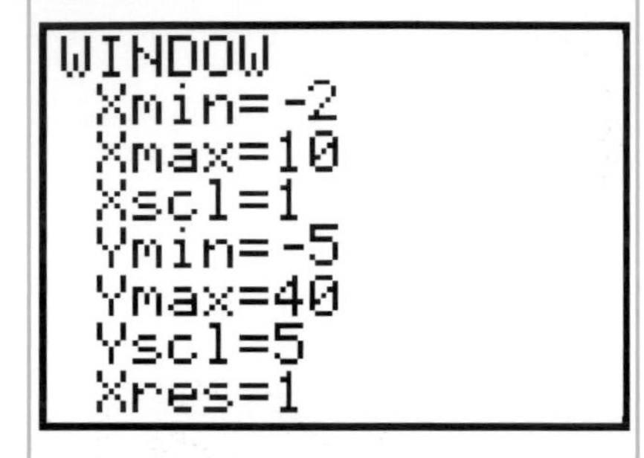

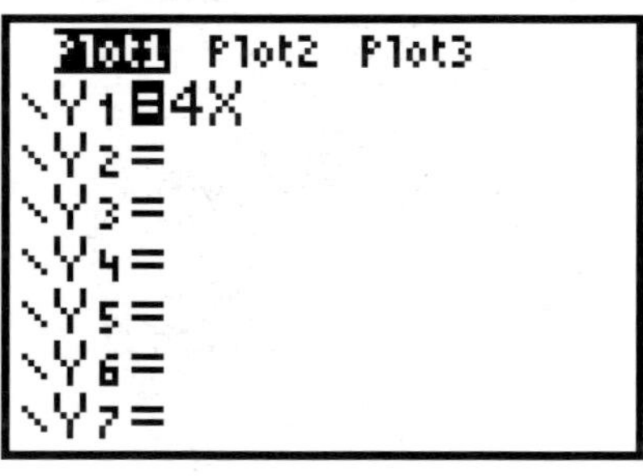

Developing the Concept of Linear Relationship

Students' prior experiences developing formulas and exploring patterns have afforded them opportunities to explore linearity informally. A formal study of linear relationships can begin with engaging students in one or more problems that connect linear relationships with real-world contexts. Stacking Cups, which involves measuring the height of a stack of cups, is one such activity (Cole and Burrill 1998, pp. 19–20).

Stacking Cups

As one teacher noted, it is a good idea to have a discussion with students about what information needs to be determined before they get started on the task.

p. 81

Goals

- Connect the concept of linearity with real-world contexts
- Use a table to organize information
- Make a graph to display data using correct labels and scales
- Recognize that a constant rate of change exists between the two variables.

Materials and Equipment

- A copy of the blackline master "Stacking Cups" for each student
- Cups of different kinds—divided into sets of cups of the same kind—that can be stacked

  You may want to have different groups of students use different types of cups. The students measure the height of one cup, two stacked cups, three stacked cups, and so on.
- Spreadsheet programs or graphing calculators (optional)

Activity

You have been hired by a company that makes all kinds of cups—foam hot cups, plastic cold cups, paper cups, and more—of different sizes. For each of the kinds of cups it makes, the company needs to know the measurements of cartons that can hold 50 cups. Your task is to provide this information.

1. Make a table and record in it the measurement data (number of cups and height of a stack of cups) for different kinds of cups.
2. Make a coordinate graph of each data set.
3. Describe what variables are being investigated and the relationship between the variables.
4. Predict how tall a stack of 50 cups would be and explain how you made your prediction.
5. For each kind of cup, recommend the inside dimensions of a carton that would hold a stack of 50 cups.
6. Compare the results for the different kinds of cups, noting similarities and differences.

Discussion

Tables 3.1 and 3.2 show two sets of data, one for eight-ounce foam hot cups, each having a diameter of 8 cm, and the other for twelve-ounce plastic cold cups, each having a diameter of 9.5 cm. It is anticipated, for this first activity, that students would make coordinate graphs like those in figure 3.3 by hand. However, either spreadsheets or graphing calculators are useful tools for students to use.

Coordinate graphs made using a spreadsheet might look like those in figure 3.3; graphs made on a graphing calculator might look like those in figure 3.4.

Students could make predictions for the height of 50 cups in a variety of ways. Looking at the table of data for the hot cups, some students might focus on the recursive relationship of adding 1.5 to the previous term to get the current term. They might use this strategy to find the height of 50 cups, possibly taking advantage of the answer function on

Fig. **3.3.**

Coordinate graphs made using a spreadsheet for the Stacking Cups activity

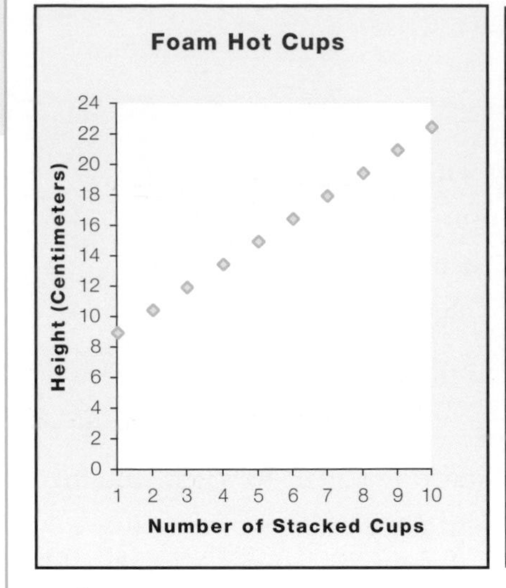

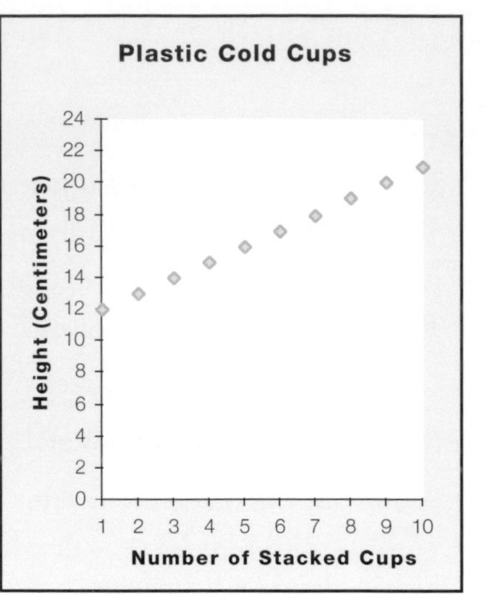

Fig. **3.4.**

Graphs for the Stacking Cups problem made using a graphing calculator

The scale for the graph in which 0 ″ *x* ″ 10 and 0 ″ *y* ″ 10*x*

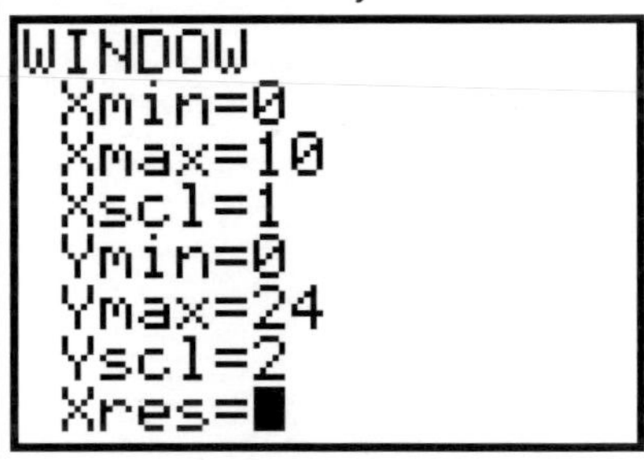

The graph for foam hot cups:

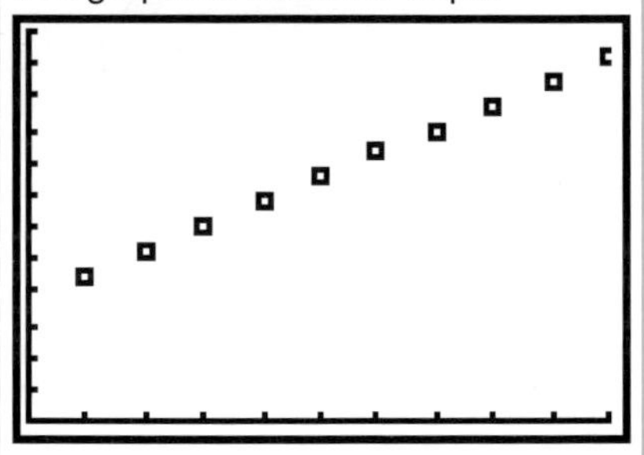

The graph for plastic cold cups:

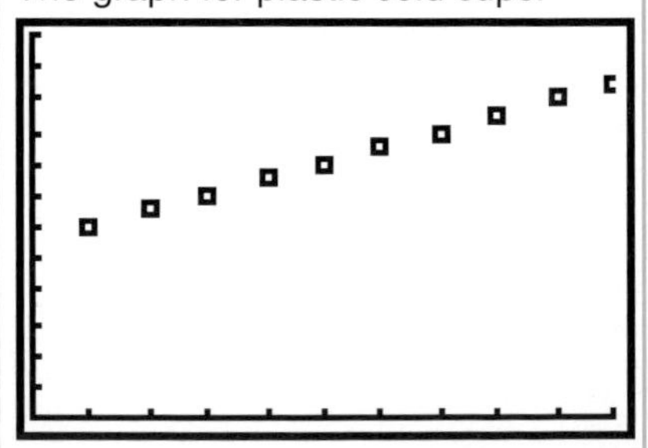

Table 3.1
Foam Hot Cups

Number of Cups	Height (cm)
1	9
2	10.5
3	12
4	13.5
5	15
6	16.5
7	18
8	19.5
9	21
10	22.5

Table 3.2
Plastic Cold Cups

Number of Cups	Height (cm)
1	12
2	13
3	14
4	15
5	16
6	17
7	18
8	19
9	20
10	21

the calculator (see fig. 3.4) or the fill-down feature of the spreadsheet to do this work for them. Other students might observe a functional relationship between the height of the first cup and the heights of the additional cups as they are stacked. The first hot cup, for example, is 9 cm in height; the 2-cup stack is 10.5 cm in height, or 1(1.5 cm) + 9 cm, and so on. A 50-cup stack, then, would be 49(1.5 cm) + 9 cm, or 82.5 cm, in height. A rule written in words might be expressed this way:

> Take the number of stacked cups and subtract 1. Multiply this result by 1.5 cm, and then add the height of the original cup—9 cm. After the first cup, only 1.5 cm of the height of any other cups is visible in the stack of cups.

Students might also observe that as the stack of hot cups grows, it becomes taller than the same number of cold cups. Initially, the first cold cup was taller than the first hot cup, but because the lip of the hot cups is wider, the stacked hot cups grow in height more quickly than the stacked cold cups do.

One carton made to store the hot cups could contain all 50 cups in a single stack; possible measurements might be h = 83 cm, w = 8.5 cm, and l = 8.5 cm. This set of measures gives "wiggle room" for the cups to fit in the carton. Students may observe that this carton is rather tall;

another carton could contain two stacks of 25 cups side by side. For this choice, the height of 25 cups—45 cm—must be determined. The carton's dimensions might be h = 45.5 cm, w = 17 cm, l = 8.5 cm, again permitting wiggle room.

Developing the Concept of Constant Rate, or Slope, and the Concept of *y*-Intercept and Exploring their Interactions

Problems like Stacking Cups direct students' attention to the idea of constant rate of change as it might occur in real-world contexts. It is important to connect these ideas more formally with the concept of linearity. One way to do so is to explore problems like Walking Rates that focus explicitly on rate. In an earlier activity, Walking Strides, students focused on the time it took them to walk a given distance using strides of different sizes. Here, the task (adapted from Lappan et al. 1998a, pp. 15–20) is modified.

Walking Rates

Goals

- Connect the concept of linearity with real-world contexts
- Use a table to organize information
- Make a graph to display data, using correct labels and scales
- Recognize the relationship among the table, the graph, and the equation and the slope of the line

Materials and Equipment

- A copy of the blackline master "Walking Rates" for each student

p. 82

Activity

Distribute copies of the blackline master, and have the students complete the tasks for the following problem presented on it:

> Several students are planning to participate in a walkathon to help raise money for charity. The distance to be walked is 10 kilometers. The students are wondering how long it might take them to walk this distance. Several students decide to do an experiment to determine their walking rates. Here are data for three of the students:

Name	Walking Rate
Jeff	1 meter per second
Rachel	1.5 meters per second
Annie	2 meters per second

> From these walking rates, a table can be organized that shows the distance walked by each of the students after various numbers of seconds (see fig. 3.5).

Fig. **3.5.**

A table for displaying the distances walked by students for certain times

Time in Seconds	Distance in Meters		
	Jeff	Rachel	Annie
0	0	0	0
1	1	1.5	2
2			
3			

1. Complete the table for several times. (Fig. 3.6 shows a table for the data.)
2. On one coordinate grid, make a graph of the time and distance data for the three people. (Figure 3.6 shows a graph for the time and distance data.) Describe how the walking rates affect the graphs. (The faster people walk, the steeper the graph becomes.)
3. For each of the three people, describe the relationship between the time and the distance walked, using words. Write an equation for each relationship, using d to represent distance in meters and t to represent time in seconds. Describe how the walking rate affects the equation.

   (Equations: Jeff, $d = t$; Rachel, $d = 1.5t$; Annie, $d = 2t$. The rate at which d increases is determined by the walking rate.)

Time (Seconds)	Distance (Meters)		
	Jeff	Rachel	Annie
0	0	0	0
1	1	1.5	2
2	2	3	4
3	3	4.5	6
4	4	6	8
5	5	7.5	10
6	6	9	12
7	7	10.5	14
8	8	12	16
9	9	13.5	18
10	10	15	20

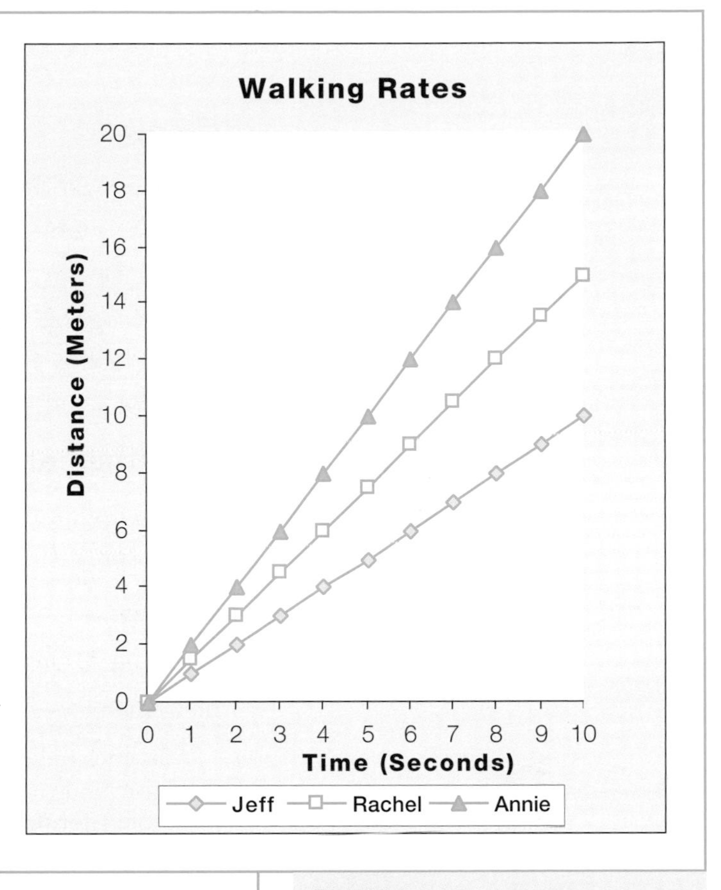

Fig. **3.6.**

A table and a graph for the time and distance data for Walking Rates

4. Using these equations, determine the distance traveled in 1 minute, 30 minutes, and 1 hour. Use this information to estimate how much time it will take for each person to complete the walkathon. (Distance traveled in 1 min: Jeff, 60 m; Rachel, 90 m; Annie, 120 m. Distance traveled in 30 min: Jeff, 1800 m [1.8 km]; Rachel, 2700 m [2.7 km]; Annie, 3600 m [3.6 km]. Distance traveled in 1 h: Jeff, 3600 m [3.6 km]; Rachel, 5400 m [5.4 km]; Annie, 7200 m [7.2 km]. Time to complete the 10-km walkathon: Jeff, 2 h, 46 min, 40 s; Rachel, 1 h, 51 min, 7 s; Annie, 1 h, 23 min, 20 s. These predictions assume that they can maintain their respective paces over long periods of time.)

Discussion

This problem focuses attention on the relationship among the table, the graph, and the equation and the slope of a line. This activity may be extended in Pledge Plans in order to introduce the concept of *y*-intercept.

Pledge Plans

Goals

- Connect the concept of linearity with real-world contexts
- Use a table to organize information
- Make a graph to display data using correct labels and scales
- Recognize the relationship among the table, the graph, the equation, and the slope of the line
- Identify the *y*-intercept from a graph or a table

Materials and Equipment

p. 83

- A copy of the blackline master "Pledge Plans" for each student
- A spreadsheet program or a graphing calculator (optional)

Activity

Distribute copies of the blackline master, and have the students complete the tasks for the following problem presented on it:

> Several students who are participating in a 10-kilometer walkathon to raise money for charity need to decide on a plan for sponsers to pledge money for the walkathon. Jeff thinks that \$1.50 per kilometer would be an appropriate pledge. Rachel suggests \$2.50 per kilometer because it would bring in more money. Annie says that if they ask for too much money, people won't agree to be sponsors; she suggests that they ask for a donation of \$4.00 and then \$0.75 per kilometer.

1. Make a table showing the amount of money a sponsor would owe under each of the pledge plans. (Fig. 3.7 shows a table for the data.)
2. On one coordinate grid, make a graph for each of the three pledge plans. Use a dashed line to connect the points. (Fig. 3.7 shows a graph for the pledge-plan data.)
3. For each of the three pledge plans, use words to describe the relationship between the money earned and the distance walked. (Jeff's plan earns \$1.50 per km walked, or $m = 1.5d$. Rachel's plan earns \$2.50 per km walked, or $m = 2.5d$. Annie's plan earns \$4.00 up front and then \$0.75 per km walked, or $m = 0.75d + 4.00$.)
4. Using m to represent the money owed, write an equation that can be used to compute the money owed under each pledge plan, given the distance the student walks. Use d to represent distance. (Equations: Jeff, $m = 1.5d$; Rachel, $m = 2.5d$; Annie, $m = 0.75d + 4.00$.)
5. Describe how increasing the amount of the pledge per kilometer affects the table, the graphs, and the equations. (Increasing the amount of the pledge per kilometer increases the amount of money earned more quickly, which is shown by the steepness of the graphs.)
6. Describe what is different about Annie's plan. What happens in the table, the graph, and the equation when Annie's plan is introduced? (Annie's plan is different because she starts out with a fixed amount that is pledged, no matter what distance is traveled. The graph and table begin at \$4.00 and then increase by \$0.75 for each kilometer walked.

	A	B	C	D
1	Distance	Money Owed (Dollars)		
2	1–10 km	Jeff	Rachel	Annie
3	0	0.00	0.00	4.00
4	1	1.50	2.50	4.75
5	2	3.00	5.00	5.50
6	3	4.50	7.50	6.25
7	4	6.00	10.00	7.00
8	5	7.50	12.50	7.75
9	6	9.00	15.00	8.50
10	7	10.50	17.50	9.25
11	8	12.00	20.00	10.00
12	9	13.50	22.50	10.75
13	10	15.00	25.00	11.50

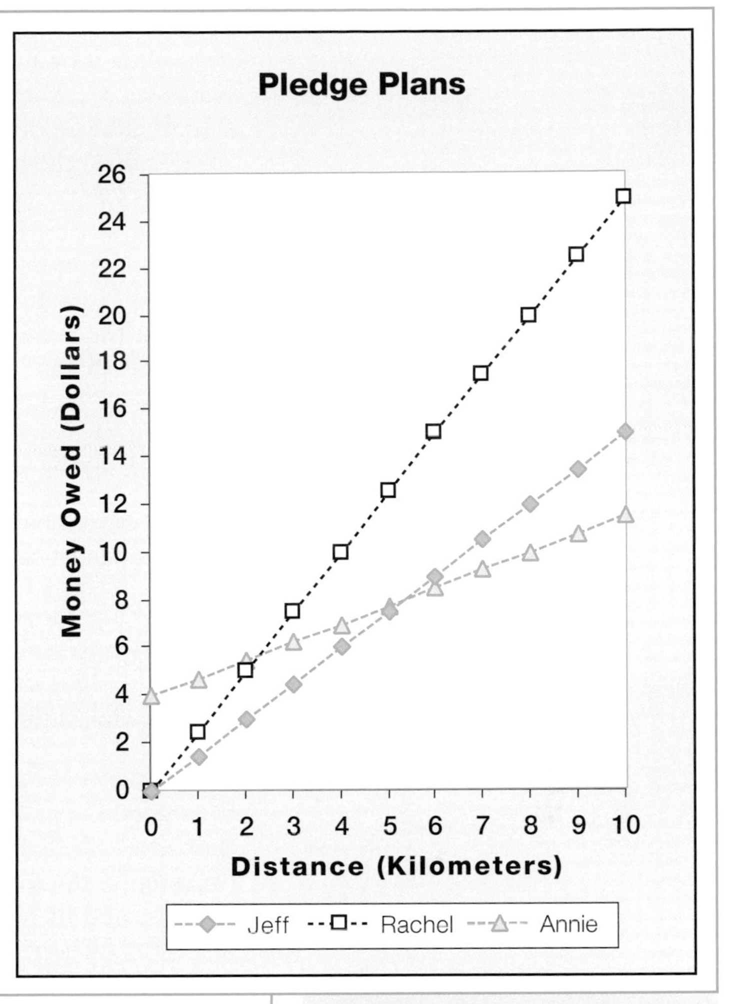

Fig. **3.7.**

A table and a graph for the money owed under the three schemes in Pledge Plans

The equation $m = 0.75d + 4.00$ indicates that the graph passes through the point (0, 4.00). The graphs of the other two plans pass through the origin, (0, 0).)

Discussion

Problems such as this one direct students' attention to various aspects of linearity, particularly the relationships among tables, graphs, and equations, and the importance of slope and y-intercept. Clearly, such problems are only a beginning to students' work in linear functions.

The concept of slope may be introduced informally by considering what it feels like to walk up different hills: Is it harder to walk up a hill with a slight incline or to walk up a hill with a steep incline? Students can draw pictures of both types of hills and analyze, for some fixed horizontal distance, the change that occurs in the vertical distance for each hill. This change gives some idea of the differences in slopes; the greater the distance of the vertical change for a fixed horizontal change, the steeper the hill.

The concept of slope can be formally developed by having students explore staircases. In comparing the steepness of different staircases, we are likely to have to consider different horizontal distances (the run or tread of a stair), unlike in the example of the steepness of a hill, in which the horizontal distances were the same (see fig. 3.8). As in the

How is the steepness of staircases related to the steepness of hills?

previous example, however, the vertical distances will be different. To determine how steep a staircase is, instead of just looking at the vertical distances, we compute the ratio of the rise to the run (or the tread) for a step. The ratio of the rise to the run for a staircase provides a measure of steepness: the larger the number, the greater the steepness of the stair. So for staircase B in figure 3.8, the ratio of the rise to the run is 2, and for staircase C, the ratio is 1/4, or 0.25. Staircase B is steeper than staircase C mathematically as well as visually!

To build staircases that are easy to climb, carpenters use the guideline that the ratio of the rise to the run for each step should be between 0.45 and 0.60 and the rise plus the run for each step should be between 17 and 17.5 inches, or approximately the length of the typical human stride. Students can do a number of different activities that focus their attention on the rise and run and the steepness of staircases:

- Measure several staircases, taking note of the rise and the run and comparing these results with those noted earlier.
- Use a computer drawing program to make pictures of staircases drawn to scale that show the relationship between rise and run that creates comfortable climbing steps. Students can experiment with drawing different variations of staircases.

The method for finding the steepness of stairs suggests a way to find the steepness, or slope, of a line drawn on a coordinate grid. From staircase C in figure 3.8, it can be seen that a line drawn from the bottom step to the top step of a staircase touches each step at one point. The rise and the run of a step are the vertical and horizontal changes, respectively, between two points on the line. Relating this idea to a line on a graph, we can talk about the steepness of the line as the *slope* according to the following definition:

$$\text{Slope} = \frac{\text{Vertical change}}{\text{Horizontal change}}$$

We can characterize lines that slant upward from left to right as having a *positive slope* (like walking up a staircase) and lines that slope downward from left to right as having a *negative slope* (like walking down a staircase). As the horizontal change is measured from left to right, the vertical change in each case is either positive or negative. Going up is a positive vertical change, and going down is a negative vertical change.

The concepts of slope and y-intercept can be explored in any number of problems. For example, consider the activity Fund Raising (adapted from Hadley [2000]).

Fig. **3.8.**
Staircases

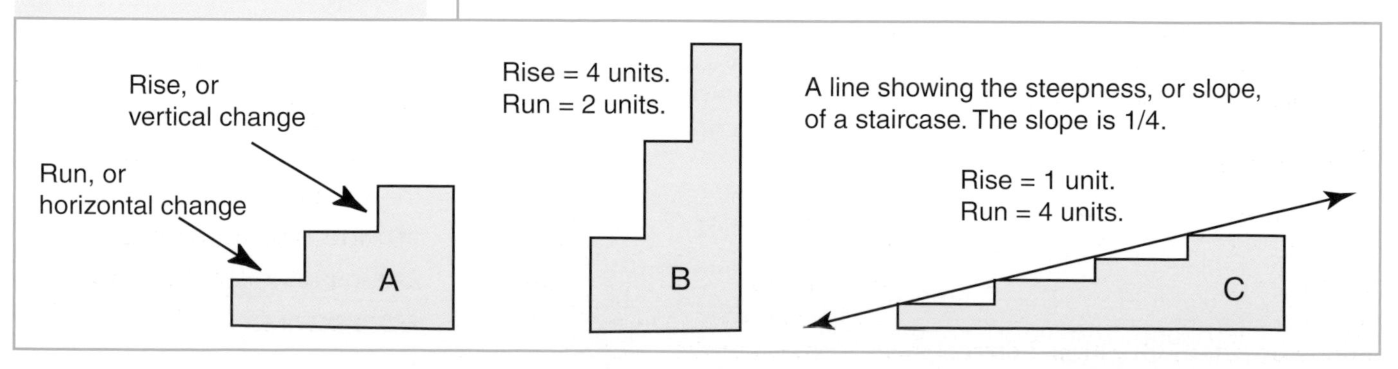

Fund Raising

Goals

- Connect the concept of linearity with real-world contexts
- Identify ways that the table, the graph, and the equation provide information to solve the problem
- Explore relationships among lines, slopes, and y-intercepts

Materials and Equipment

- A copy of the blackline master "Fund Raising"

p. 84

Activity

Distribute copies of the blackline master, and have the students complete the tasks for the following problem presented on it:

> The Middle Grades Math Club needed to raise money for its spring trip to Washington, D.C., and for its Valentine's Day dance. The club members decided to have a car wash to raise money for these projects. The club treasurer provided the following information:
>
> - The total cost of sponges, rags, soap, buckets, and other materials needed will be $117.50.
> - The usual charge for washing one car is $4.00.
>
> As a member of the club, your job is to produce a mathematical analysis of this situation that includes making a table and a graph to answer the following questions:
>
> 1. How much money would the club raise if the members washed 0, 1, 2, 3, or 4 cars? 10 cars? 50 cars? 100 cars? How can you use your table to answer these questions? How can you use your graph to answer these questions?
> 2. (*a*) If no cars are washed, what happens to the club's finances? How is this situation shown on the graph? On the table?
>   (*b*) With every car washed, what is the change in the amount of money earned? How is this change shown on the graph? On the table?
>   (*c*) Write a general rule to use to determine how much money is raised for any number of cars. How is the information you identified for parts (*a*) and (*b*) related to the rule you wrote?
> 3. How many cars must the members wash to "break even"?
> 4. How many cars must the members wash to raise enough money for—
>   - the trip to Washington—estimated cost, $875,
>   - the Valentine's Day dance—estimated cost, $350?
> 5. Realistically, can the car wash raise enough money for both of these activities? Explain why or why not.

This problem can be represented graphically using a graphing calculator or spreadsheet. Using a graphing calculator, students need to determine a formula ($y = 4x - 117.50$) for the problem. Once they have entered it, they can explore the table for the values (see fig. 3.9).

The graph reflects the table, showing 100 cars washed (see fig. 3.10). Because of the dimensions of the viewing window, the graph appears asaline although the data are discrete.

Discussion

In answering question 1, students develop both the table and the graph on the basis of the information requested. (Note, however, that the fund-raising scenario involves discrete rather than continuous values and that students'graphs should show dashed rather than solid lines if done by hand.) Question 2 directs their attention to the constant

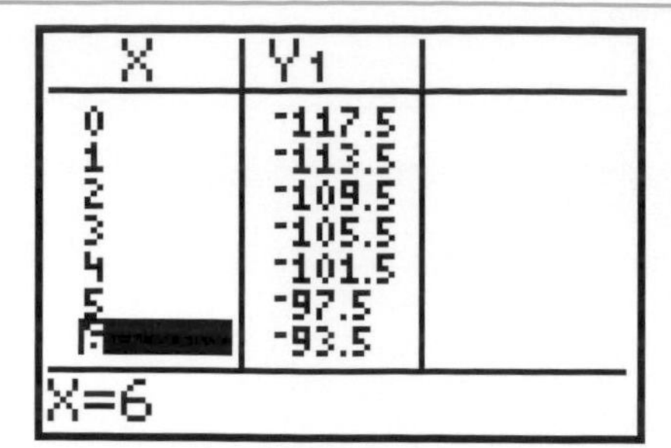

X	Y1
0	-117.5
1	-113.5
2	-109.5
3	-105.5
4	-101.5
5	-97.5
6	-93.5

X=6

The table by ones

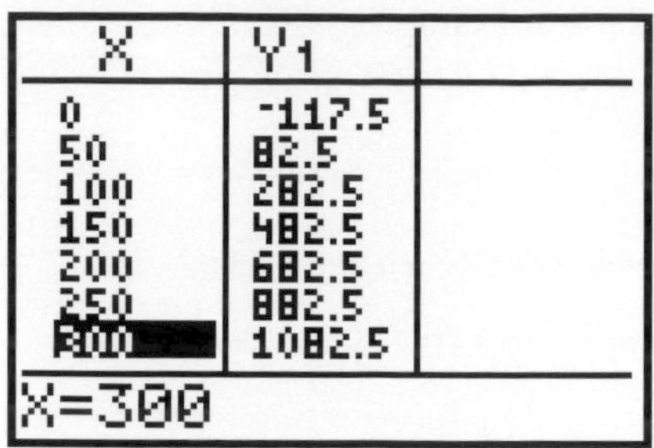

X	Y1
0	-117.5
50	82.5
100	282.5
150	482.5
200	682.5
250	882.5
300	1082.5

X=300

The table by fifties

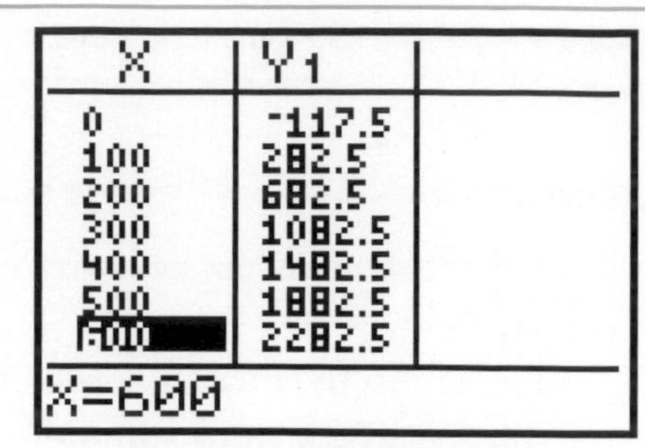

X	Y1
0	-117.5
100	282.5
200	682.5
300	1082.5
400	1482.5
500	1882.5
600	2282.5

X=600

The table by hundreds

Fig. **3.9.**

A table of values for the Fund Raising activity

The student treated each part of this problem independently, so she included the thirty car washings required to break even in the totals for each part.

change (slope), $4.00, and the y-intercept (–$117.50); these values are what are used to develop the general rule for any number of cars washed. The break-even point occurs as the students move from 29 to 30 cars washed; 30 cars washed yields a profit of $2.50. The math club needs to raise $1225, so they need to wash at least 336 cars. One student explained her reasoning thus:

> Let me first say that it is realistically possible to make enough money for the Valentine's Day dance. It will take 117 cars, but at approximately 10 or more cars a day, it can be done in about 12 days. The trip cost of $875 to Washington, D.C., can be made if 249 cars are washed. At approximately 10 or more cars a day, it can be done in 25 days. This means a lot of car washing! Maybe it would be a good idea to find other ways to raise money along with the car washes. Also, realistically, we may need to buy more sponges, rags, soap, and other car-washing materials.

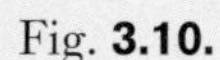

Once students have explored concepts in problem situations such as the Fund Raising activity just described, they can consider the relationships between lines, slopes, and y-intercepts, using context-free tasks like the following, which will help build a solid understanding of linearity.

1. For a linear equation (e.g., $y = 2x$ or $y = 2 - 3x$), choose any two points and compute the ratio of the vertical change to the horizontal change from one point to the other.
2. Given two points (e.g., (2, 6) and (0, 4)), plot the points on a coordinate grid and draw the line that passes through them.
   (*a*) Determine the slope of the line.
   (*b*) Mark and label at least three other points on this line.
   (*c*) Make a table using these points. How does the pattern in the table relate to the slope of the line?
   (*d*) Locate the y-intercept for the line.
3. How can you use the slope of a line to determine whether a line slants

Fig. **3.10.**

Graphs representing the Fund Raising activity. The break-even point is between $x = 29$ and $x = 30$.

Graph scale

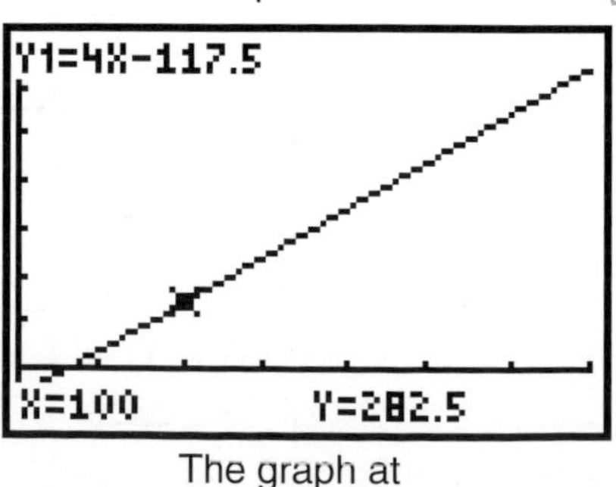

The graph at number of cars washed = 100

The graph at number of cars washed = 29

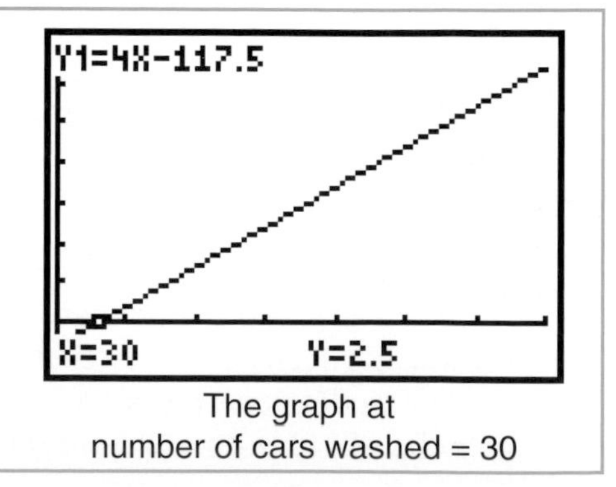

The graph at number of cars washed = 30

downward or upward from left to right? For example, one point on a line is (0, 4); name a slope that would create a line that slants downward. Graph the line using the point given and the slope you chose. Do the same for a line that slants upward.

4. If you are given two points on a line, explain how to write an equation for the line and identify both the slope and the y-intercept. For example, a line passes through (2, 7) and (6, 15). Write its equation.
5. If you are given the slope of a line and the y-intercept or a second point—for example, a line with slope 2/3 and y-intercept at (0, 2)—can you write an equation and sketch a graph for the line that meets the given conditions? Why is it necessary to be given a point? Why isn't knowing the slope sufficient to let you write an equation for a line?
6. Make a graph for each of these equations on the same coordinate grid: $y = x + 1$, $y = x + 3$, $y = x - 2$, and $y = x + 6$. What do these lines have in common? Why is this so?
7. Make a graph for each of these equations on the same coordinate grid: $y = 2x + 1$, $y = x + 1$, and $y = -3x + 1$. What do these lines have in common? Why is this so?

Solutions to Tasks

1. For $y = 2x$, two points are (1, 2) and (3, 6). The ratio of vertical change to horizontal change is
$$\frac{(6-2)}{(3-1)} = \frac{4}{2} = 2.$$
2. See figure 3.11. (*a*) The slope is 1. (*c*) The difference in the y-values from one point to another is the same as the difference in the corresponding x-values. Since slope is
$$\frac{\text{difference in } y\text{-values}}{\text{difference in } x\text{-values}},$$
the slope is 1. (*d*) The y-intercept is (0, 4).
The equation is $y = x + 4$.
3. Lines with negative slopes slant downward from left to right; lines with positive slopes slant upward from left to right. A line slanting downward with a slope of –3 would pass through points (0, 4) and (1, 1).
A line slanting upward with slope of 3 would pass through points (0, 4) and (1, 7).
4. We can determine the slope from the given points. Since
$$m = \frac{(15-7)}{(6-2)} = \frac{8}{4} = 2,$$
the slope is 2. Since y decreases by 2 whenever x decreases by 1, the y-intercept is (0, 3). So the equation of the line is $y = 2x + 3$.
5. A line with slope 2/3 means
$$\frac{\text{vertical change}}{\text{horizontal change}} = \frac{2}{3}.$$
In other words, if the x-coordinate increases by 3, the y-coordinate increases by 2. Since (0, 2) is one point on the line, we can use the slope to find other points, such as (3, 4), (6, 6), and (9, 8).
The equation is $y = (2/3)x + 2$.
6. See the graph in figure 3.12. The lines are parallel because they have the same slope. The y-intercepts are different.
7. See the graph in figure 3.13. The lines intersect at (0, 1). The slopes are different. The y-intercepts are the same.

Fig. **3.11**

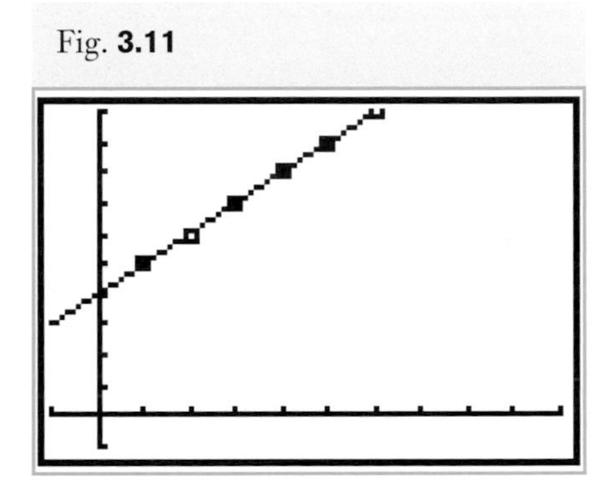

Fig. **3.12.**

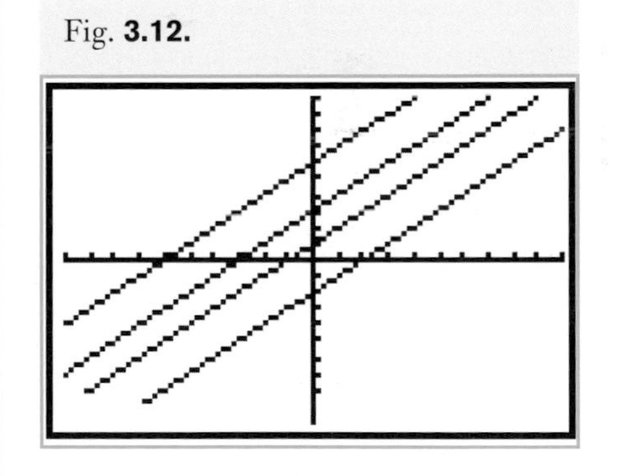

Fig. **3.13.**

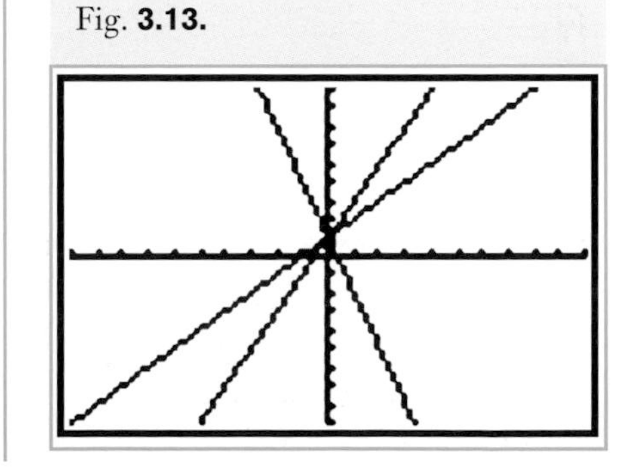

Contrasting Linear Relationships with Other Kinds of Relationships

We want students to be able to look at a table or at a graph of a linear function and identify it as representing a linear relationship. They should be able to identify a constant rate of change, which is the slope and results in a line as the graph of the relationship. When linear functions are compared to quadratic functions or exponential functions, the differences in the tabular, the graphical, and the symbolic representations make the characteristics of linearity apparent. Consider these three problems and the applicable tables, graphs, and equations.

Spreadsheets lend themselves to the use recursive strategies for describing relationships. See figure 3.14 for a table and a graph of the two savings plans.

Shoe-box plan: With the first term = 15, new term = old term + 15.

Uncle's plan: With the first term = 15, new term = old term + 0.05 × old term + 15.

1. Jason has a dog-walking business from which he earns \$30 each month. Each month Jason saves \$15 from the money he has earned. Consider these two situations:

   (*a*) Each month, Jason saves the \$15 in a shoe box under his bed. At the end of a year, how much money has Jason saved?

   (*b*) Each month, Jason gives the \$15 to his uncle, who pays him 5% interest each month on the accumulated balance, which includes all interest earned to date. At the end of a year, how much money has Jason saved?

   Make tables and graphs (on the same grid) to show each situation. For this problem, connect the points with dashed lines to highlight the pattern in the data. Examine both graphs. How are they alike? Different? Examine both tables. Describe the rate of change for the shoe-box plan and for Jason's uncle's plan. How are they different? One situation may be characterized as being linear. Which situation is linear? How can you tell?

2. Squares are special kinds of rectangles for which all sides are the same length. To find the perimeter of a square, we add the lengths of the four sides or multiply the side length by 4 (4*s*). To find the area of a square,

Fig. **3.14.**

A table and graph for Jason's two savings plans

	A	B	C
1	Month	Shoe-Box Plan	Uncle's Plan
2	1	15	15.00
3	2	30	30.75
4	3	45	47.29
5	4	60	64.65
6	5	75	82.88
7	6	90	102.03
8	7	105	122.13
9	8	120	143.24
10	9	135	165.40
11	10	150	188.67
12	11	165	213.10
13	12	180	238.76

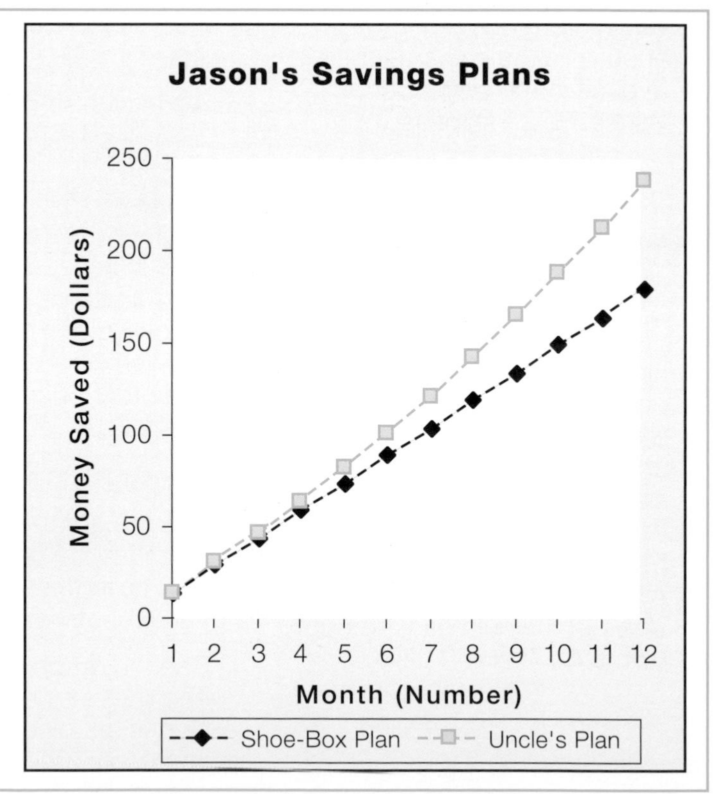

we multiply the length by the width or, because the side lengths are the same, we can describe this multiplication as side × side, or s^2. Explore what happens to the perimeters and areas of a square as you increase and decrease the length of the side. Organize your information in tables, and make graphs and connect the points in order to see the patterns. How do the tables and graphs compare? What can you say about the rate of change in each of the two situations?

It makes sense to use an explicit strategy for describing each of the two relationships in this problem. See figure 3.15 for a table and a graph of the perimeters and areas of squares. From the table and graph, we can discover the following:

$$\text{Perimeter} = 4 \times \text{side}.$$
$$\text{Area} = \text{side} \times \text{side, or side}^2.$$

3. My neighbor, Sal, is an avid gardener. He wants to enclose a garden plot in a field using part of an existing fence and 32 feet of fencing that he has purchased. What is the largest rectangular garden plot he can make using part of the existing fence as one side of the rectangle? What is the area of the garden plot? What is the length of the fence parallel to the existing fence? Use tables, graphs, and diagrams to help you solve this problem. When you have finished, discuss how the tables and graphs that show the area and the length of the opposite fence compare. What can you say about the rate of change in each of the two situations? Which situation is linear? How do you know?

It makes sense in this problem to use an explicit strategy for describing each of the two relationships. Figure 3.16 shows a table and graphs for garden plots. From the table and graphs, we can discover the following:

Fig. **3.15.**

A table and a graph for the perimeters and areas of squares

Side Length	Perimeter	Area
1	4	1
2	8	4
3	12	9
4	16	16
5	20	25
6	24	36
7	28	49
8	32	64
9	36	81
10	40	100
11	44	121
12	48	144
13	52	169
14	56	196
15	60	225
16	64	256
17	68	289
18	72	324
19	76	361
20	80	400

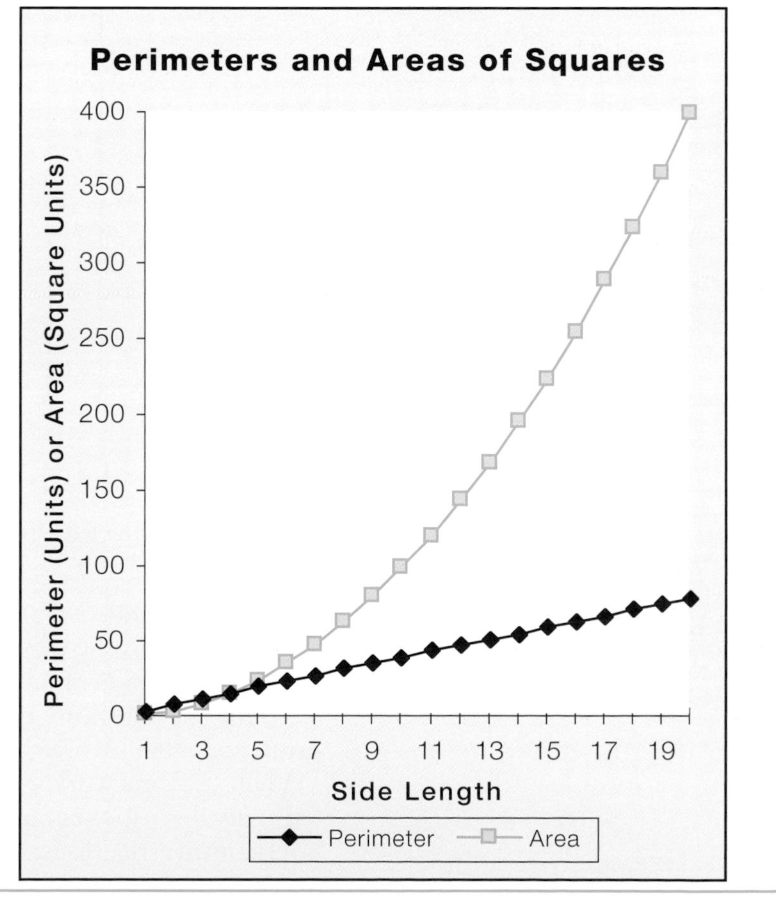

Connected Side Length	Opposite Side Length	Area of Plot
1	30	30
2	28	56
3	26	78
4	24	96
5	22	110
6	20	120
7	18	126
8	16	128
9	14	126
10	12	120
11	10	110
12	8	96
13	6	78
14	4	56
15	2	30

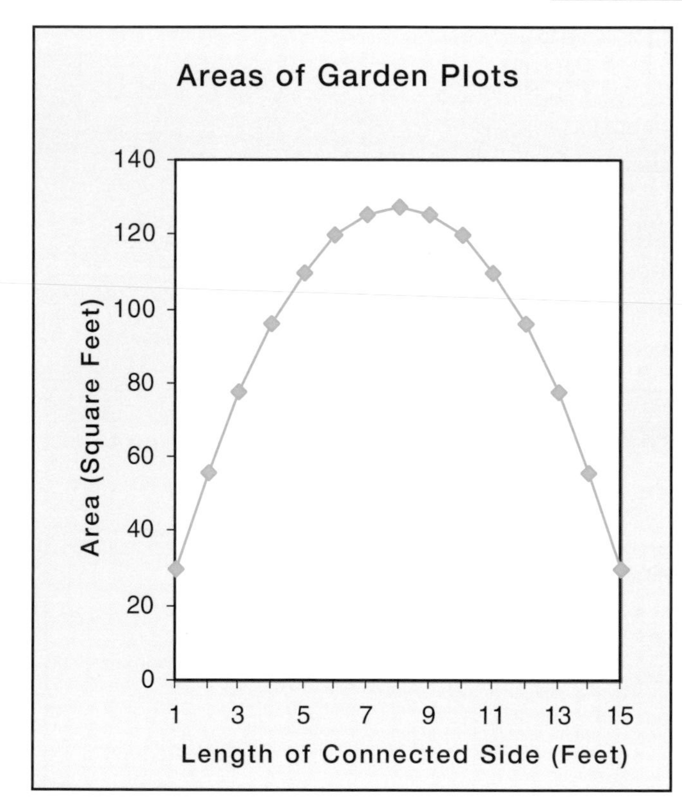

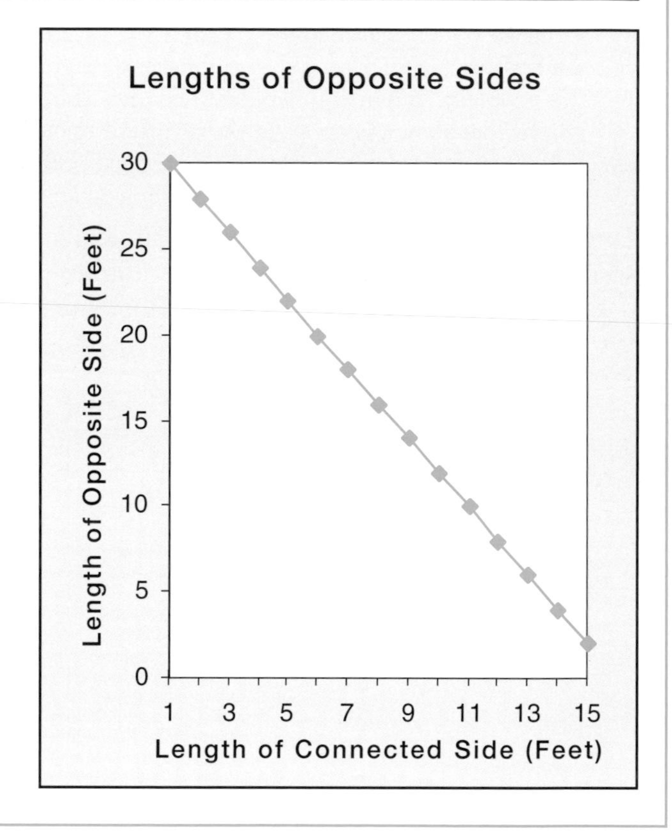

Fig. **3.16.**

A table and graphs for the areas and side lengths of rectangular garden plots

The length of the opposite side = 32 – 2 × (length of the connected side).

The area of the garden plot = (length of the connected side) × (length of the opposite side).

Each rectangle is built off the existing fence so that two "connecting sides" attach to the existing fence and then to the opposite side. Two possible fenced areas are shown in figure 3.17.

This section shows the tables, graphs, and rules for three different situations that involve linear and quadratic or exponential functions. When students work with linear functions, examples of problems that also involve exponential or quadratic functions can be included as part of their investigations. As students explore these different functions, the

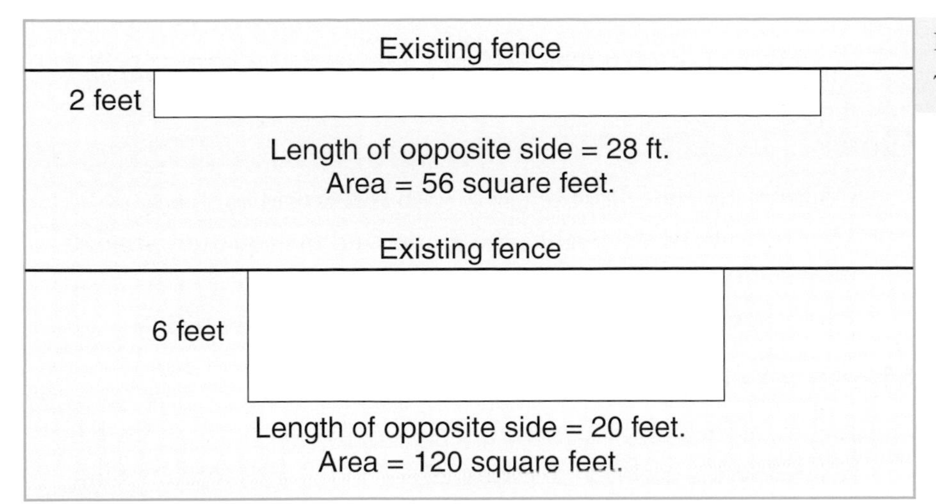

Fig. **3.17.**

Two possible fenced areas

characteristics that distinguish linear functions from the other types of functions can be highlighted. Students can compare the tables. What do they notice about the rates of change in linear models compared with those in exponential or quadratic models? In linear models, the change involves the repeated addition of a constant quantity. With the other two models, the change involves repeated multiplication. The graphs of linear models are lines; the graphs of the other types of functions are curves.

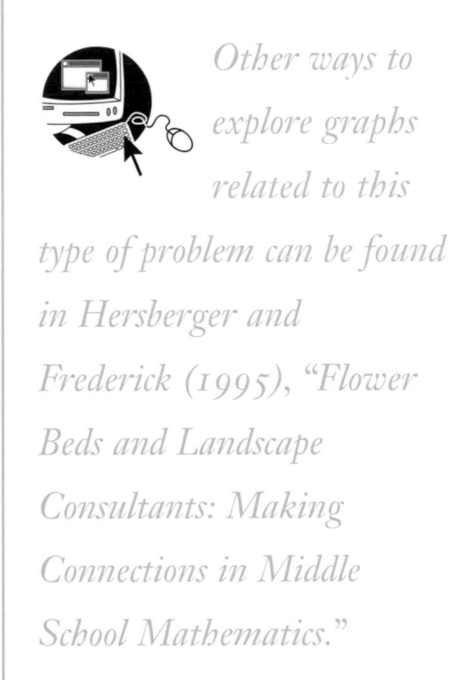

Solving Linear Equations

Work with linear functions often involves comparing two or more situations and then using graphs and tables to comment on characteristics of the situations as they relate to one another and identifying points of intersection as needed. It may involve writing equations as well; at this point, however, symbolic solutions are not the focus.

Printing Books is an activity that students could explore after working with simpler conditions posed in other problem contexts (e.g., see the discussion of cellular telephone pricing plans in *Principles and Standards for School Mathematics* [NCTM 2000, pp. 223–25, 229–31]).

Printing Books

Goals

- Connect the concept of linearity with real-world contexts
- Identify ways that the table, the graph, and the equation provide information to solve the problem
- Explore relationships among lines, slopes, and y-intercepts

p. 85

Materials and Equipment

- A copy of the blackline master "Printing Books" for each student
- A spreadsheet program to create tables and graphs (optional but useful)

Activity

Have the students complete the tasks for the following problem, which is presented on the blackline master:

> Purchasing decisions for an algebra textbook for the school system must be made. The Board of Education needs to know the costs of providing a 325-page algebra textbook to three different schools. The schools have agreed to pilot a new textbook this year; the publisher is making page proofs of the book available, but the schools must make the copies needed for the students. An assistant has done some research and has discovered the following three possibilities:
>
> 1. A local printing company: The algebra textbook can be printed by a local printer for a cost of $9.50 per book, with an initial cost of $5000 for typesetting.
> 2. A local copy center: The algebra textbook can be duplicated at a local copying center for $0.05 per page plus $2.00 per book for binding.
> 3. The school district: The school district's own copying center can reproduce the textbook at a cost of $0.035 per page plus an up-front cost of $3000.
>
> In the past, the algebra textbook orders have never exceeded 2250 books and the cost has never exceeded $35,000. Using tables and graphs, as well as identifying general rules, slopes, y-intercepts, and points of intersection, do a mathematical analysis of these different options and write a recommendation for the Board of Education to consider.

In this problem, it makes sense to use an explicit strategy for describing each of the two relationships. See figures 3.18 and 3.19.

Printing Company: Cost = 9.50 × (number of books) + 5000.

Copy Center: Cost = (0.05 × 325 + 2) × (number of books).

School District: Cost = (0.035 × 325) × (number of books) + 3000.

Discussion

This problem involves a number of different tasks. First, tables and graphs for each situation need to be made. Using a spreadsheet makes this problem much easier; see figures 3.18 and 3.19. Figure 3.18 shows the costs of the three plans for up to 2500 books. The copy center's plan will be the most expensive, and the printing company's, the least expensive if all 2250 books are produced at one location. However, suppose that these additional conditions apply:

- Western High School will need 400 textbooks next year.

This activity has been adapted from Hadley (2000).

	A	B	C	D
1	No. of Books	Printing Co.	Copy Center	School District
2	0	5000.00	0.00	3000.00
3	100	5950.00	1825.00	4137.50
4	200	6900.00	3650.00	5275.00
5	300	7850.00	5475.00	6412.50
6	400	8800.00	7300.00	7550.00
7	500	9750.00	9125.00	8687.50
8	600	10700.00	10950.00	9825.00
9	700	11650.00	12775.00	10962.50
10	800	12600.00	14600.00	12100.00
11	900	13550.00	16425.00	13237.50
12	1000	14500.00	18250.00	14375.00
13	1100	15450.00	20075.00	15512.50
14	1200	16400.00	21900.00	16650.00
15	1300	17350.00	23725.00	17787.50
16	1400	18300.00	25550.00	18925.00
17	1500	19250.00	27375.00	20062.50
18	1600	20200.00	29200.00	21200.00
19	1700	21150.00	31025.00	22337.50
20	1800	22100.00	32850.00	23475.00
21	1900	23050.00	34675.00	24612.50
22	2000	24000.00	36500.00	25750.00
23	2100	24950.00	38325.00	26887.50
24	2200	25900.00	40150.00	28025.00
25	2300	26850.00	41975.00	29162.50
26	2400	27800.00	43800.00	30300.00
27	2500	28750.00	45625.00	31437.50

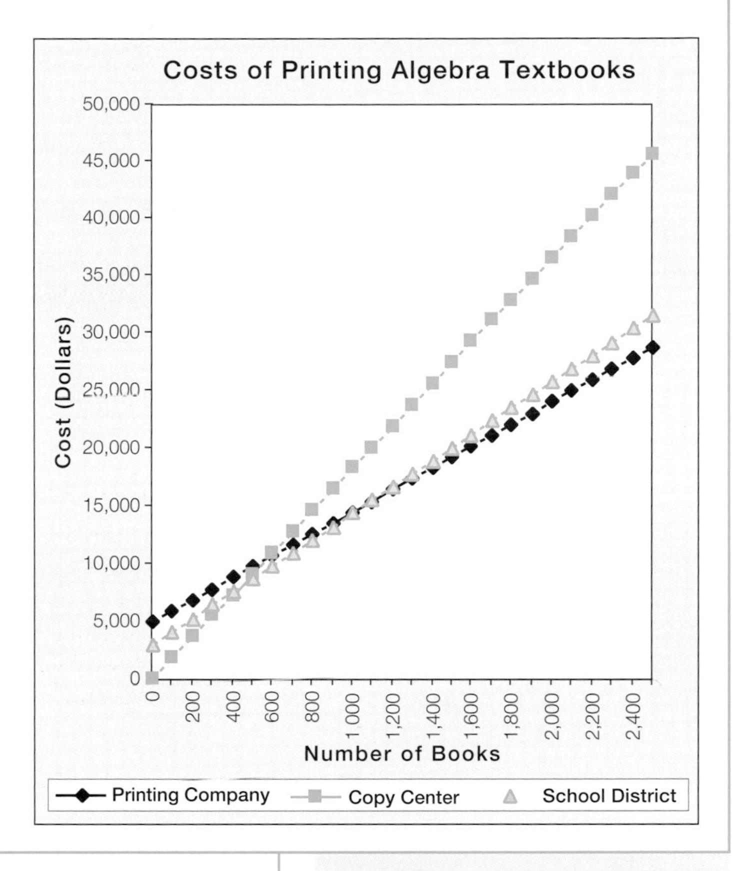

Fig. **3.18.**

A table and a graph representating the costs of three printing plans

- Eastern High School will need 550 textbooks next year.
- Northern High School will need 1400 textbooks next year.

If each high school can use a different printing company, which choice should each make? Figure 3.19 displays additional data and a graph on which the zoom feature allows us to see more clearly where the costs for different plans intersect. On the basis of these data, one student wrote the following:

> Western High School will need 400 textbooks for next year, so the cheapest way of having these books made would be to use the local copy center. It would cost $7,300 dollars for these books. Eastern High School needs 550 books. The cheapest place to go would be the school district's in-house copy center. It would cost $9,256.25. Northern High School needs 1400 books. The cheapest way to get these books would be to go with the printing company. This would cost $18,300. All together, these three orders would cost $34,856.25. If all 2350 books were produced by one company, the cheapest choice would be the printing company. This would cost $27,325. It actually would cost less to produce all the books together, rather than letting the individual schools order their texts.

	A	B	C	D
1	No. of Books	Printing Co.	Copy Center	School District
2	0	5000.00	0.00	3000.00
3	25	5237.50	456.25	3284.38
4	50	5475.00	912.50	3568.75
5	75	5712.50	1368.75	3853.13
6	100	5950.00	1825.00	4137.50
7	125	6187.50	2281.25	4421.88
8	150	6425.00	2737.50	4706.25
9	175	6662.50	3193.75	4990.63
10	200	6900.00	3650.00	5275.00
11	225	7137.50	4106.25	5559.38
12	250	7375.00	4562.50	5843.75
13	275	7612.50	5018.75	6128.13
14	300	7850.00	5475.00	6412.50
15	325	8087.50	5931.25	6696.88
16	350	8325.00	6387.50	6981.25
17	375	8562.50	6843.75	7265.63
18	400	8800.00	7300.00	7550.00
19	***425***	9037.50	***7756.25***	***7834.38***
20	***450***	9275.00	***8212.50***	***8118.75***
21	475	9512.50	8668.75	8403.13
22	500	9750.00	9125.00	8687.50
23	525	9987.50	9581.25	8971.88
24	550	10225.00	10037.50	9256.25
25	***575***	***10462.50***	***10493.75***	9540.63
26	600	10700.00	10950.00	9825.00
27	625	10937.50	11406.25	10109.38

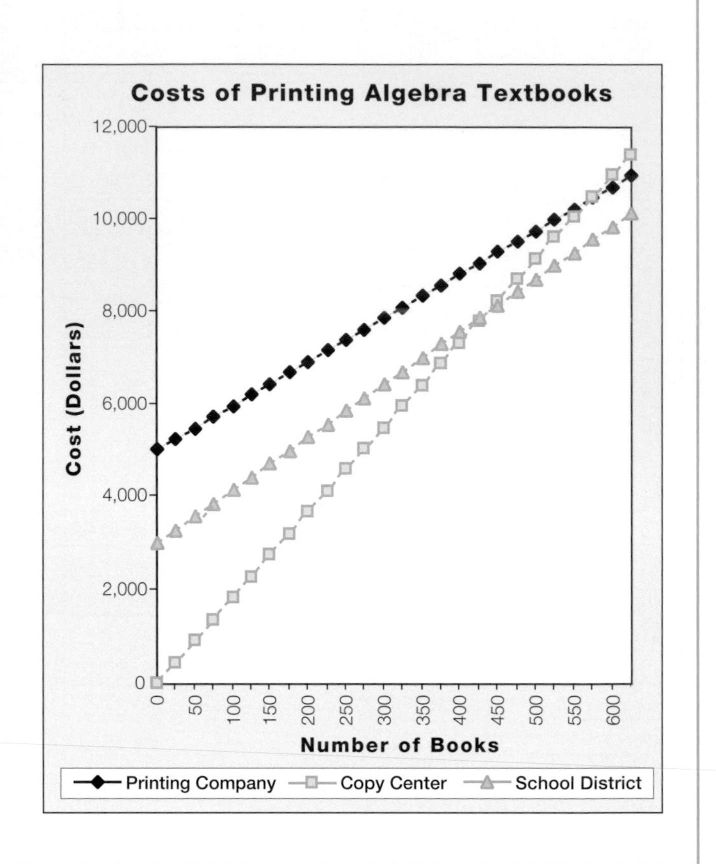

Fig. **3.19.**

A more detailed table and a zoomed-in view of the graph representing three printing plans

GRADES 6–8

NAVIGATING *through* ALGEBRA

Chapter 4
Using Algebraic Symbols

Algebraic formula

Graph

Table

Concrete or pictorial representation

Verbal description

According to *Principles and Standards for School Mathematics* (NCTM 2000), students' understanding of symbolic representations and operations used in algebra should emerge from explorations of problems set in a variety of contexts that embody the concept of function. In the middle grades, greater emphasis is placed on the relationships among the five representations of a problem. Interpreting, writing, evaluating, and operating using symbolic expressions develop from and after extensive experiences with the other kinds of representations and their use in modeling functional patterns.

Important Mathematical Ideas

Learning some basic ideas and skills of reasoning with purely symbolic expressions opens new and powerful ways of understanding quantitative change. Yet these ideas need to be developed in ways that allow students to continue to make sense of and connect the underlying concepts. Problems that raise the essential ideas in meaningful settings serve to introduce conventional algebraic notation and techniques. Once such ideas have been developed in a context, it is possible to generalize the explorations. The following are some basic ideas for middle-grades students to understand:

- In many situations, characterizing the relationships among the variables can result in a number of equivalent expressions; being able to determine when expressions are equivalent is an important part of algebra.

For further ideas, see Korala and Goodwin (2000) and O'Connor (2000).

When you first use this puzzle, ask the students to choose whole numbers. Then have them create a spreadsheet (see fig. 4.1b) so they can easily experiment with different numbers. Once they have their spreadsheets completed, suggest that they explore the problem, experimenting with different types of numbers, for example, −6 or 202.4 or 0.34. Is the final result 4 when calculated on the spreadsheet? What happens if the students work the problems mentally, without the spreadsheet? What does it mean for a number to be "1 less than" −6 or 202.4 or 0.34? What does the formula used in the spreadsheet suggest as an answer to this question? What challenges are hidden in this problem?

- The order in which mathematical operations are performed is important; the convention for the order of operations is a guideline for carrying out operations.
- The distributive property relates multiplication and addition and expanding or factoring an expression as needed; it is of central importance to the study of algebra.
- The commutative properties (or rearrangement properties) for addition and multiplication in an algebra context are applied to symbolic representations.
- In solving linear equations symbolically, we use the strategy of undoing the mathematical operations until the variable is isolated on one side of the equation. While doing so, we are careful to apply the mathematical operations to both sides of the equation to maintain the equivalence of expressions.

What Might Students Already Know about These Ideas?

Students will have had a number of algebraic experiences prior to formal work with symbols. Many of these experiences engage them in generalizing and in writing algebraic expressions to describe patterns or to generalize solutions to problems presented in context. One way to move students to more-formal work with symbols and to assess their fluency with such work is to explore number puzzles. Such problems lend themselves to the use of variables in more-abstract contexts. Consider the following "think of a number" puzzle:

- Write down any whole number.
- Add the number that is 1 less than the original number.
- Add 9 to this result.
- Divide the sum by 2.
- Subtract the original number.
- What is the final result?

Students can perform the steps as directed, first choosing whole numbers, and determine their results. Each person's final result will be 4. To explore this puzzle further, a spreadsheet can be displayed, showing several entries and results for this activity (see fig. 4.1a). Students may suggest other whole numbers that can be entered as replacements for any of the initial numbers shown (e.g., 5, 18, 10, and so on). The spreadsheet will automatically calculate the results—4 each time!

The task for students is to write the spreadsheet formulas for the cells that will reproduce this number puzzle. One way to think about this problem is to acknowledge that only one variable is involved—the initial number that is selected. However, when spreadsheets are used, each step in the puzzle may actually generate a new variable in the sense that the spreadsheet "rule" for that given cell often names the contents of the prior cell as the variable that is input to the computation needed for the current cell. For example, in cell B7 of the spreadsheet in figure

	A	B	C	D	E	F
1						
2	1. Write down any whole number.	5	18	10	76	100
3						
4	2. Add the number that is	9	35	19	151	199
5	1 less than the original number.					
6						
7	3. Add 9 to this result.	18	44	28	160	208
8						
9	4. Divide the sum by 2.	9	22	14	80	104
10						
11	5. Subtract the original number.	4	4	4	4	4

(a)

	A	B	C	D	E	F	G	H	I
1							Decimal	Decimal	Integer
2	1. Write down any number.	5	18	10	76	100	202.4	0.34	–6
3									
4	2. Add the number that is	B2 + B2 –1	35	19	151	199	403.8	–0.32	–13
5	1 less than the original number.								
6									
7	3. Add 9 to this result.	B4+ 9	44	28	160	208	412.8	8.68	–4
8									
9	4. Divide the sum by 2.	B7/2	22	14	80	104	206.4	4.34	–2
10									
11	5. Subtract the original number.	B9 – B 2	4	4	4	4	4	4	4

(b)

Fig. **4.1.**

Spreadsheets for a number puzzle

4.1b, we see the formula is "B4 + 9," which commands the spreadsheet to add 9 to the value in cell B4 and enter the sum in cell B7. It would also be possible to enter the formula "B2 + B2 – 1 + 9" in cell B4, which would yield the same results. Continuing to use spreadsheet notation, we could express the final result (in cell B11) for column B using only the original input variable in the formula as (((2 * B2) – 1 + 9)/2) – B2, which is equivalent to the result of 4. In this example, all the cell formulas are written in terms of the input variable and not in terms of the interim computed variables that occur as each step in the problem is completed. This experience with spreadsheets requires that students understand what the variables are, the relationships between the steps in the puzzle, and how variables may be used to characterize these relationships.

When this activity is used for preassessment, the teacher needs to be concerned with students' facility with, and ease of use of, variables to describe the relationships among the steps in the problem. With spreadsheets, it is important to see if students use a recursive strategy, defining the current cell on the basis of the contents of the previous cell and the operation to be carried out on the results found in that cell in order to determine the value of the current cell. You might also notice if any students define all the cells on the basis of the cell (the variable in the problem) that holds the original number. It is worthwhile to have a discussion of both strategies.

Students' fluency and confidence in describing how to generalize the steps in a think-of-a-number puzzle provide insights into their readiness to move to more-formal work with symbols, which is the traditional notion of algebra.

This problem can be explored without the use of spreadsheets as well by naming the original number n and having students describe what happens to n. Here is one way this might occur:

Step 1: n

Step 2: $n + n - 1 = 2n - 1$

Step 3: $2n - 1 + 9 = 2n + 8$

Step 4: $(2n + 8) \div 2 = n + 4$

Step 5: $n + 4 - n = 4$

Selected Instructional Activities

The activities in this chapter are intended to highlight some important components of algebra that need to be developed symbolically. These activities are examples of ways that symbolic work can grow out of meaningful situations and that the symbolism can emerge from the context.

Exploring Equivalent Expressions

Many students' explorations of problems expose them to equivalent expressions. The Tiling Tubs problem is an activity that can focus students' attention on the equivalence of expressions.

Tiling Tubs

Goals

- Write equations to describe the relationships among variables
- Determine when expressions are equivalent

Materials and Equipment

- A copy of the blackline master "Tiling Tubs" for each student

p. 86

Activity

Have the students complete the following tasks, which are presented on the blackline master:

> Hot tubs and in-ground swimming pools are sometimes surrounded by borders of tiles.
>
> 1. How many 1-foot square tiles will be needed for the border of a square hot tub that has edge length *s* feet?
> 2. Express the total number of tiles in as many ways as you can.
> 3. Be prepared to convince your classmates that the expressions are equivalent.

Discussion

Samples of students' responses to this problem are shown in figure 4.2. Unlike problems in which students may have looked at a pattern of change as pools increase in size, in this task, students must be able to state the relationship that is determined by knowing a variable side length, *s*, and the context of the problem. The results can be represented by a variety of equivalent expressions. Students can discuss ways to determine whether one expression is equivalent to another. For

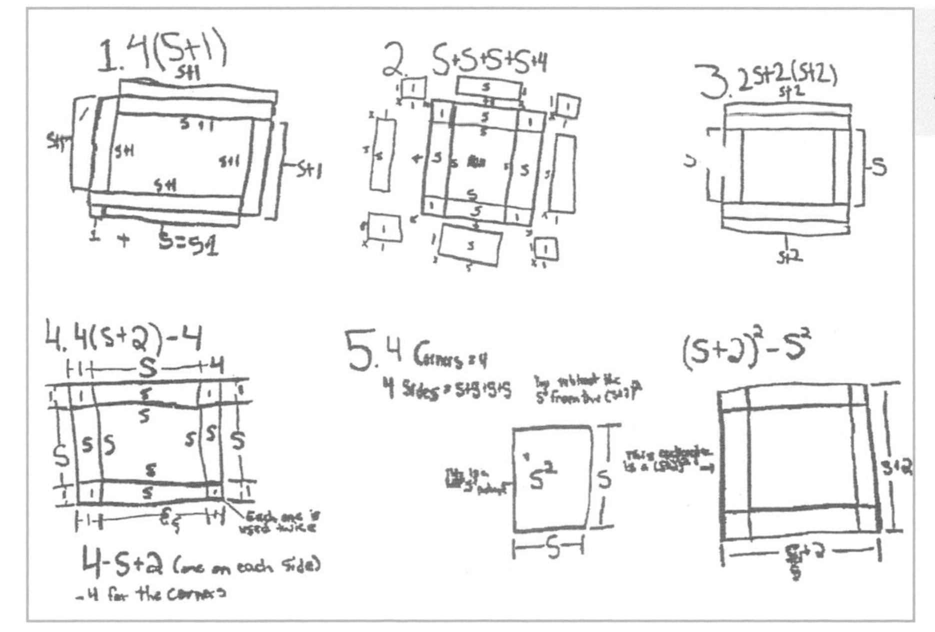

Fig. **4.2.**

Students' solution strategies for the Tiling Tubs problem

This activity has been adapted from Lappan et al. (1998b, pp. 20–22).

September 1998

		1	2	3	4	5
6	7	8	9	10	11	12
13	14	15	16	17	18	19
20	21	22	23	24	25	26
27	28	29	30			

example, in examining whether $4(s + 1)$ is equivalent to $s + s + s + s + 4$, one student noted that

> If we have $4(s + 1)$, that is the same as $(s + 1)$ four times, or $(s + 1) + (s + 1) + (s + 1) + (s + 1)$. When I add up all the 1s, I can see that it is the same as $s + s + s + s + 4$.

Other contexts may involve situations that are not geometric, such as the think-of-a-number puzzles or the following calendar puzzle:

> Select a month from a calendar. Select any 4-by-4 grid of 16 days. Draw a square around your 4-by-4 grid.
>
> 1. Find the sum of the dates in the two diagonals of your grid. (72 in the example shown)
> 2. Find the sum of of the dates in the four corners of your grid. (72 in this example)
> 3. Find the sum of the four numbers in the center square inside your grid. (72 in this example)
> 4. What do you observe about the three sums? Can you explain what is happening?

A natural part of students' looking back at their solutions will be determining why the sums are equal in all three cases, no matter which 4-by-4 grid they select. Since the grids may vary, you can suggest calling the first number in the grid a variable. The students should agree on a letter to represent the first number in their grids. Ask them to name each of the other numbers in the grid. For example, if 6 is represented by D (day), 14 is represented by $D + 8$, 22 is represented by $D + 16$, and 30 is represented by $D + 24$, so the sum of this diagonal is $D + (D + 8) + (D + 16) + (D + 24)$. The sum of the other diagonal is $(D + 3) + (D + 9) + (D + 15) + (D + 21)$. Both sums are equivalent to $4D + 48$, which also equals the sum of the four corner numbers, as well as the sum of the four inside-square numbers. Students may want to explore other 4-by-4 grids on the calendar for the same month or for other months. If they do, you can ask questions such as "If D is the smallest number in the grid, what is the relationship of D to the remaining cells? Why is this so?"

Exploring the Use of the Commutative, Associative, and Distributive Properties and the Order of Operations

In the middle grades, mathematical tasks emphasize the speed and ease of using the commutative, associative, and distributive properties in conjunction with the order of operations. We assume that students are familiar with the commutative and associative properties; if not, these properties should be discussed briefly. In addition, students may need time to explore the distributive property before using it to evaluate expressions. If so, then problems such as Plotting Land help students understand this property.

Plotting Land

Goals

- Write equations to describe the relationships among variables
- Explore the use of the distributive property

Materials and Equipment

- A copy of the blackline master "Plotting Land" for each student

p. 87

Activity

Distribute copies of the blackline master and have the students complete the following tasks:

> A gardener has divided a plot of land into two sections. The shaded section is to be used for growing vegetables, and the unshaded portion is for growing flowers.
>
> 1. What is the area of the shaded portion of the plot of land?
> 2. What is the area of the unshaded portion of the plot of land?
> 3. What is the area of the entire plot of land?
> 4. Suppose the gardener wanted to divide the plot of land so that 3/5 of the land was used for vegetables and 2/5 of the land was used for flowers.
>   (*a*) What would be the dimensions of the shaded and unshaded portions?
>   (*b*) What would be the areas of the shaded and unshaded portions of the plot of land?
>   (*c*) What would be the area of the entire plot of land?

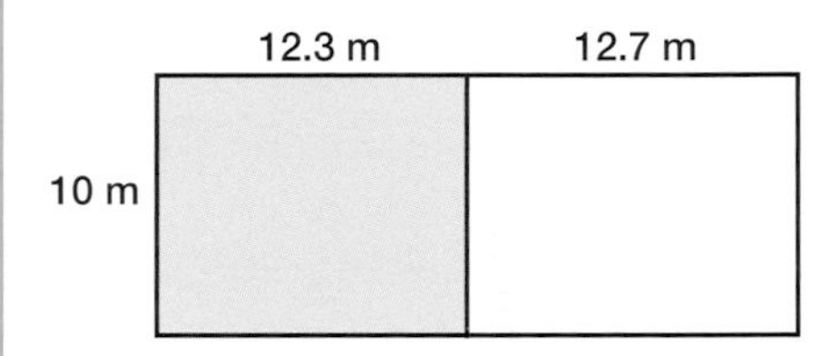

Discussion

Teachers can use this problem to help students focus on the equivalence of two expressions. For parts 1–3, the equivalence is

$$(10 \times 12.3) + (10 \times 12.7) = 10(12.3 + 12.7).$$

For parts 4a–4c, the equivalence is

$$(10 \times 15) + (10 \times 10) = 10(15 + 10).$$

Although the areas of the two sections of the plot of land vary, the area of the entire plot of land remains constant. Activities like this one help students become familiar with the distributive property as they work with numbers.

Instruction in the order of operations can promote students' awareness and use of the commutative, associative, and distributive properties. By designing problems that focus attention on the value of these properties in simplifying expressions so they can be evaluated, teachers build students' understanding before they work formally with these properties in algebra. Consider each of the following problems and the value of using the commutative, associative, and distributive properties along with the order of operations:

Using the associative property

$$-12 + (-88 + 59) = (-12 + -88) + 59 = -100 + 59 = -41$$

Using the distributive property

$$8 \times 6\frac{3}{4} = 8 \times \left(6 + \frac{3}{4}\right) = (8 \times 6) + \left(8 \times \frac{3}{4}\right) = 48 + 6 = 54$$

$$17 \times 51 + 17 \times 49 = 17(51 + 49) = 17 \times 100 = 1700$$

Using the commutative property

$$\frac{2}{3} + \frac{3}{5} + \frac{1}{3} + \frac{2}{5} = \frac{2}{3} + \frac{1}{3} + \frac{3}{5} + \frac{2}{5} = 1 + 1 = 2$$

The order of operations and the properties are important in understanding work in algebra. A variety of experiences can be designed that help students focus on algebraic applications of these concepts; only one is considered here. Figure 4.3 provides information about pricing plans for two different video stores.

Fig. **4.3.**

Pricing plans for Video Busters and Video Arts stores

Many people like to rent videotapes of films to watch at home. Often, they go to the video rental store to find new releases—videos of many of the films that have just finished playing in the movie theaters. New releases can be quite popular, so video stores often have different, higher rental rates for new releases. When the releases are just out (brand new), the stores limit the number of days that they can be borrowed. Generally, after a year, videos that continue to be popular are called *favorites;* favorites have constant, lower rental rates. Consider the pricing plans of these two stores:

Video Busters

Type of Video	Price for Each Video Rented
Brand-new releases (0 to 2 months old)	$3.99 for 2 days
New releases (more than 2 months but less than 12 months old)	$3.99 for 5 days
Favorites (one year or older)	$2.99 for 5 days

Video Arts

Type of Video	Price for Each Video Rented
Brand-new releases (0 to 1 month old)	$3.50 for 1 day
New releases (more than 1 month but less than 12 months old)	$3.50 for 3 days
Favorites (one year or older)	$2.00 for 3 days

In solving any problems involving these pricing schemes, we will assume that each month has 30 days.

Once the students have read the problem, a number of different questions can be asked that focus on understanding the problem context, that involve order of operations and properties, and that lead students to explore variables. These questions include the following:

1. What is the maximum possible income earned by Video Busters in the first two months of renting a new release? By Video Arts? Show several different ways you could determine the answers to the questions.

Answers will vary. An example:

Video Busters

$3.99 × (30 ÷ 2) + $3.99 × (30 ÷ 2) = $119.70

Video Arts

$3.50 × (30) + $3.50 × (30 ÷ 3) = $140

2. What is the maximum possible income earned by Video Busters in the first year of renting a new release? By Video Arts? Show several different ways you could determine the answers to the questions.

Answers will vary. An example:

Video Busters

$\$3.99 \times (30 \div 2) \times 2 + \$3.99 \times (30 \div 5) \times 10 = \$359.10$

Video Arts

$\$3.50 \times 30 + \$3.50 \times ((30 \div 3) \times 11) = \$490$

3. After the first year of availability of a new release, the income earned is constant, assuming that the video is fully rented each month.

   (*a*) For Video Busters, the total income earned in month 1 of the second year is ($\$2.99 \times 30/5$), or \$17.94. At the end of month 2, the cumulative second-year income is $\$17.94 \times 2$, or \$35.88, and so on. Write a rule for determining the cumulative income from a favorite video in the second and subsequent years if you know how many months it has been available during those years. Let x equal the number of months.

   Money earned = $\$17.94x$

   (*b*) For Video Arts, the income earned in month 1 of the second year is ($\$2.00 \times 30/3$), or \$20.00. At the end of month 2, the cumulative second-year income is ($\$2.00 \times 60/3$), or \$40.00, and so on. Write a rule for determining the cumulative income from a favorite video in the second and subsequent years if you know how many months it has been available during those years. Let x equal the number of months.

   Money earned = $\$20.00x$

4. My friend created the spreadsheet in figure 4.4 to show costs for renting up to ten videos from Video Arts. Together, we rented some brand-new releases and some favorites from Video Arts. We spent more than \$15.00 and less than \$20.00. What are the possible combinations of brand-new videos and favorites that we could have rented? Show each possibility as a statement like the following: 3 brand new + 3 favorites = \$16.50.

	A	B	C	D
1	Video Arts		Cost	
2	No. of Videos	Brand New	New	Favorites
3	1	3.50	3.50	2.00
4	2	7.00	7.00	4.00
5	3	10.50	10.50	6.00
6	4	14.00	14.00	8.00
7	5	17.50	17.50	10.00
8	6	21.00	21.00	12.00
9	7	24.50	24.50	14.00
10	8	28.00	28.00	16.00
11	9	31.50	31.50	18.00
12	10	35.00	35.00	20.00

Fig. **4.4.**

A spreadsheet showing the costs of renting brand-new and favorite videos from Video Arts

Solving Linear Equations

Solving linear equations may be approached in a variety of ways. Initially, it is helpful to demonstrate the need to solve equations by using realistic contexts. Continuing with a video context, consider this problem (adapted from Cai and Kenney 2000, pp. 536–37).

> The Video Arts store has made some modifications in its rental plans. It now offers two annual rental plans.
>
> 1. Plan A has a $20 annual membership fee, and all videos rent for $2.00 per day.
> 2. Plan B has no membership fee, and videos rent for $2.50 per day.
>
> For what number of video rentals will these two plans cost the same? Explain your thinking.

In considering the ways that students might solve this problem, recall that by now students should be fairly fluent in using different representations. It is possible to set up a table using a spreadsheet to examine how the costs increase under both plans (see fig. 4.5). From this table, we can see that renting forty videos using either plan will cost the same amount, $100.

It also is possible to reason about the situation. For each rented video, plan B costs $0.50 more than plan A costs, so $20.00 ÷ $0.50 per video rental = 40 videos. Plan A and plan B would cost the same when forty videos are rented in a year.

Some students might solve this problem graphically. One solution is shown in figure 4.6. The data points for plans A and B intersct at forty videos at a cost of $100.

Other students might reason symbolically about this problem. Letting n equal the number of videos rented, we can write the following rules for these plans:

The cost for plan A is $2n + 20$.

The cost for plan B is $2.5n$.

Fig. **4.5.**

A spreadsheet displaying data for two video-rental plans

	A	B	C
1	No. of Videos	Plan A	Plan B
2	0	20.00	0.00
3	5	30.00	12.50
4	10	40.00	25.00
5	15	50.00	37.50
6	20	60.00	50.00
7	25	70.00	62.50
8	30	80.00	75.00
9	35	90.00	87.50
10	40	100.00	100.00
11	45	110.00	112.50
12	50	120.00	125.00

Fig. **4.6.**

A graph of the data for two video-rental plans

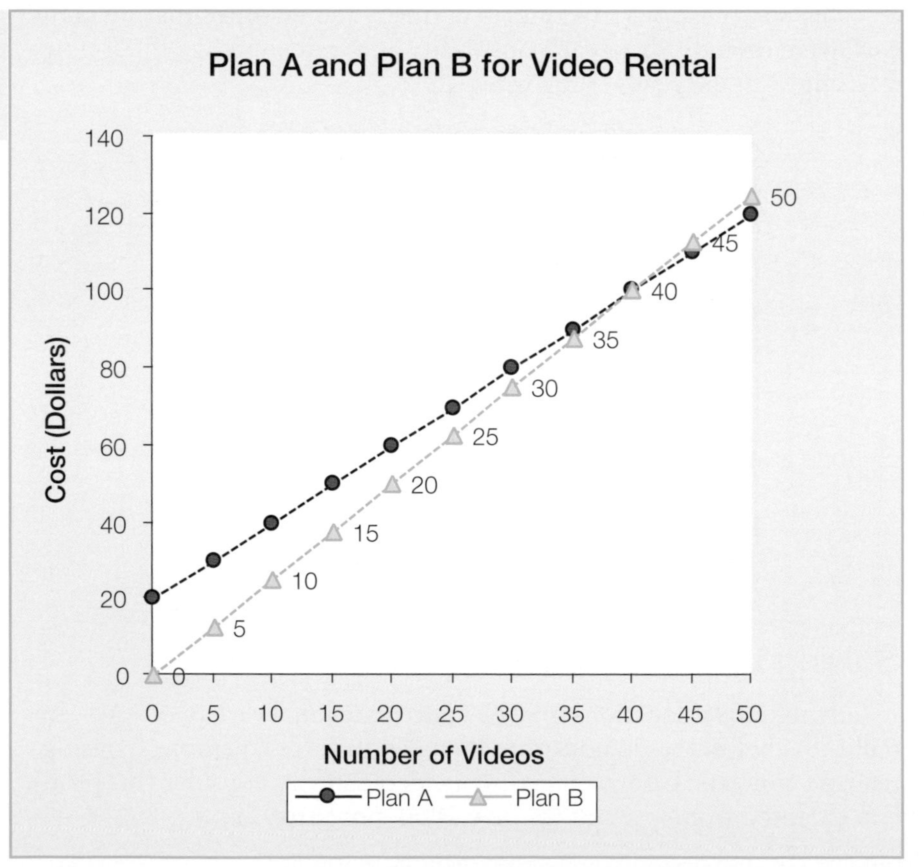

We set $2.5n$ equal to $2n + 20$. It is here that we want to begin reasoning about solving equations algebraically. Students can discuss how they might solve these two equations. They can experiment with substituting values for n until they find one that yields equivalent results. Another way is to reason with symbols. In the formula $2n + 20$, $2n$ means \$2.00 for each video rented. The expression $2.5n$ means \$2.50 for each video rented. Subtracting the variable video cost of $2n$ from $2n + 20$ leaves the membership cost of \$20. To balance this action, we need to adjust the video cost of $2.5n$ by reducing it by an equivalent amount so that we have $2.5n - 2n$, or $0.5n$. This means that for some number of videos, the \$0.50-per-video difference between the two plans is equal to \$20.00 (the original membership fee). This fee accounts for forty videos at a rate of \$0.50 per video. This reasoning is shown symbolically thus:

$$
\begin{aligned}
2.5n &= 2n + 20 \\
2.5n - 2n &= 2n - 2n + 20 \\
0.5n &= 20 \\
0.5n \div 0.5 &= 20 \div 0.5 \\
n &= 40
\end{aligned}
$$

Substituting 40 videos for n in each equation shows that when n equals 40 (the number of videos rented), the cost will be the same under either plan.

Certainly, we want students to move from reasoning within the problem context to working with more-abstract symbols and equations to solve such problems. Often this transition to equation solving is supported by an analogy to a balance scale, with the emphasis on the use of inverse operations and on keeping the scale "in balance" as changes are made (see, e.g., Billstein and Williamson [1999, p. 58]). We want students eventually to solve problems using symbolic methods and to have these methods make as much sense to them as any of the other methods.

Students can also explore multiple equations and their relationships. For example, consider the spreadsheet with several values substituted in different equations that involve the variable A (see fig. 4.7). The selected expressions highlight equivalence using the distributive property (e.g., $6 - 3A$ and $3(2 - A)$), the commutative property (e.g., $6 - 3A$ and $-3A + 6$), or selected values of A (e.g., $6 - 3A = A + 14$ when $A = -2$). This spreadsheet can emphasize the use of a table to help solve an

	A	B	C	D	E	F
1	A	6 – 3A	3(2 – A)	–3A + 6	–6A	A + 14
2	–5	21	21	21	30	9
3	–4	18	18	18	24	10
4	–3	15	15	15	18	11
5	–2	12	12	12	12	12
6	–1	9	9	9	6	13
7	0	6	6	6	0	14
8	1	3	3	3	–6	15
9	2	0	0	0	–12	16
10	3	–3	–3	–3	–18	17
11	4	–6	–6	–6	–24	18
12	5	–9	–9	–9	–30	19

Fig. **4.7.**

A spreadsheet that allows students to explore multiple equations and their relationships

equation. For example, for what value of A is $-6A = A + 14$? For what value of A is $A + 14 = 6 - 3A$? Similar questions can extend students' thinking about solving equations to solving inequalities. For example, for what values of A is $-6A < A + 14$? Students can be encouraged to ask their own questions and to create their own spreadsheets with several equations included.

GRADES 6–8

NAVIGATING *through* ALGEBRA

Looking Back and Looking Ahead

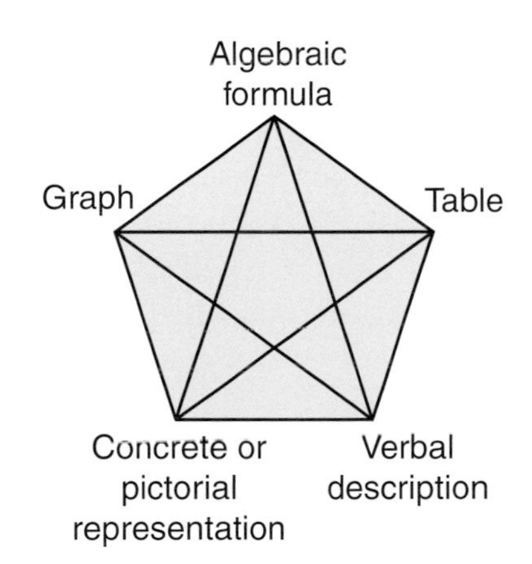

In the middle grades, work in algebraic thinking moves from informal explorations to formal considerations of algebra. The emphasis is on the use of, and relationships among, five representations of a problem (see the figure at the left). Attention is focused particularly on problem situations. Students think about algebra in context. The connections among the problem situation and the various representations occur in several different ways.

Patterns continue to provide a foundation for algebraic thinking. Growing patterns, which are developed through geometric and other contexts, are explored through the use of verbal descriptions and tables. In the example of gardens of different sizes, students verbally describe what garden 4 or garden 10 looks like, paying attention to what changes and what stays the same in the different garden areas. They organize in tables counts of the numbers of tiles needed to make each garden border and may make use of recursive patterns to identify the number of tiles needed for progressively larger gardens:

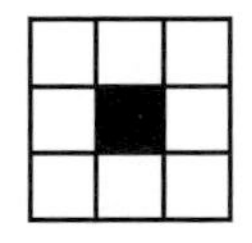
Garden 1

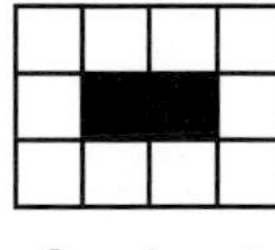
Garden 2

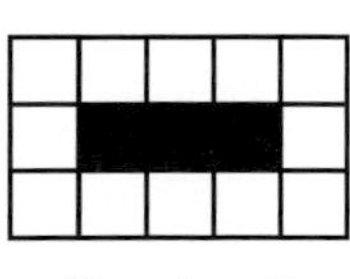
Garden 3

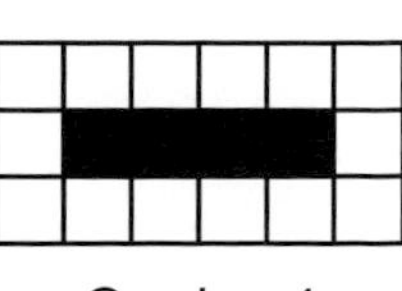
Garden 4

Writing an explicit rule to describe the number of tiles needed for any garden area in the sequence (e.g., $y = 2x + 6$) involves referring to the verbal description. The use of more-formal strategies for function definition occurs in high school.

Graphs are an essential tool for representing functions. Students need numerous opportunities to explore graphs, particularly graphs that involve changes over time in nonroutine problem situations. Such experiences help students focus on the components of a graph and what information a graph can provide about a situation. More-formal work with graphs develops as students focus on linearity.

After many experiences with different representations, students see the connections among the representations and consider ways to describe problems that can be organized and structured in a table or as a graph. Throughout, students continue to pay attention to what changes and how it changes and to what stays the same.

Linear functions are the focus of middle-grades work; it is expected that students move to high school with a solid, rich, and well-developed understanding of linearity and the relationships among representations of linear functions. Slope, as an expression of rate of change, and y-intercept, as well as an analysis of general behaviors associated with linear functions and their representations, become important topics in the curriculum. Although the attention is on linear functions, students are exposed to quadratic and exponential functions both to highlight the distinctions among the three families of functions and to give students the informal experiences they will need for a formal study of other families of functions in high school.

Algebraic symbols and operations receive formal attention in the middle grades. Such work emerges, again, from problem situations. In a problem context, students can make sense of what it means to "solve for x" by relating their solution to the problem. At the same time, they are asked to integrate their understanding and use of mathematics properties and generalized arithmetic strategies in order to reason about more-abstract symbol manipulation.

The middle grades provide a bridge that connects the informal work with algebra in the prekindergarten-to-grade-5 curriculum with the formal mathematics of the secondary school. The thoughtful, careful development of essential algebraic concepts in grades 6–8 provides the foundation for the advanced work students will encounter in grades 9–12.

Grades 6–8

Navigating *through* Algebra

Appendix

Blackline Masters

Exploring Houses

Name ______________________________

Build these houses using the orange squares and the green triangles from pattern-block pieces:

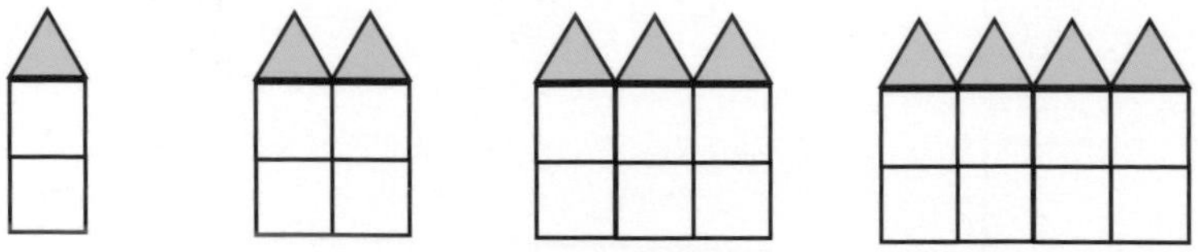

House 1 House 2 House 3 House 4

1. For each house, determine the total number of pieces needed. ______________________

   How many squares and triangles are needed for a given house? ______________________

   Organize your information in some way.

2. Describe what house 5 would look like. ______________________________________

   ____

   ____

   Draw a sketch of this house.

3. Predict the total number of pieces you will need to build house 15. ________________

   Explain your reasoning.____

   ____

   ____

4. Write a rule that gives the total number of pieces needed to build any house in this sequence.

   ____

   ____

Building with Toothpicks

Name ______________________________

The shapes shown below are made with toothpicks. Look for patterns in the number of toothpicks in the perimeter of each shape.

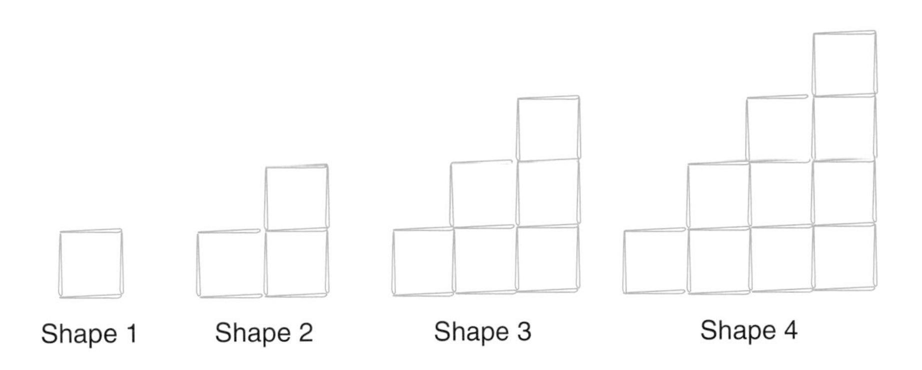

1. Use a pattern from the shapes above to determine the perimeter of the fifth shape in the sequence. Show or explain how you arrived at your answer. ______________________________

____

2. Write a formula that you could use to find the perimeter of any shape *n*. Explain how you found your formula. ______________________________

____

____

____

This activity has been adapted from Burkhardt et al. (2000)

Bouncing Tennis Balls

Name ______________________________

How many times can you bounce and catch a tennis ball in two minutes?

A bounce is defined as dropping the ball from your waist. Work with a team of four people (including yourself). One of the team members keeps the time while you bounce and catch the ball. A second team member counts the bounces, and a third team member records the data in a table showing the cumulative number of bounces. Each trial consists of a two-minute experiment, with the number of bounces recorded after every ten seconds. The timekeeper calls out the time at ten-second intervals. When the time is called, the counter calls out the number of bounces that occurred during that ten-second interval. The recorder writes this count in the table and keeps track of the cumulative number of bounces.

Time (Seconds)	Number of Bounces during Interval	Cumulative Number of Bounces
0	________	________
10	________	________
20	________	________
30	________	________
40	________	________
50	________	________
60	________	________
70	________	________
80	________	________
90	________	________
100	________	________
110	________	________
120	________	________

The same process is followed by each person on the team, with the team members rotating roles, so that each person can collect a set of data. All team members must bounce the ball on the same surface (e.g., tile, carpet, concrete) because differences in the surface could affect the number of bounces.

Once the data have been collected, use centimeter graph paper to make a coordinate graph showing your cumulative number of bounces over two minutes.

Walking Strides

Name ______________________________

Working in teams of two people, set up an experiment in which one person, the walker, walks one-quarter of a given distance (e.g., one-quarter of 80 or 100 feet). The walker chooses to use a short, regular, or long stride and uses this stride for one-quarter of the predetermined distance. A second person, the timekeeper, records the time it takes the walker to walk this distance. The walker repeats this procedure for each of the three strides. Then the walker and the timekeeper switch places and repeat the experiment.

Once a team has completed gathering the data, each person uses this information to estimate the time it would take to walk half the distance, three-quarters of the distance, and, finally, the total distance, using each of the three strides and assuming that each person's pace is constant. These estimates are also recorded.

Walking Paces (________ Total Distance)

	Time in Seconds for Different Strides			
Distance	Short	Regular	Long	
1/4 distance (_____)	_______	_______	______	(Record your actual times here.)
1/2 distance (_____)	_______	_______	______	(Estimate your times on the basis of your actual times for one-quarter of the distance.)
3/4 distance (_____)	_______	_______	______	
Total distance (_____)	_______	_______	______	

Make graphs of these data, plotting the points by hand on centimeter graph paper or using a graphing calculator.

From Stories to Graphs

Name ______________________________

1. In a walking experiment, Josephine walked a total distance of 40 feet. At the halfway point, she had walked for 25 seconds. She stopped for 5 seconds to tie her shoe and then continued walking for 25 more seconds. Sketch a graph that shows Josephine's distance from the starting point over time.

---

2. You are gathering data in the school cafeteria from 8:00 A.M. to 3:00 P.M. Sketch a graph that tells a story about the number of cans of soda in a vending machine over that time. Write a paragraph that tells the same story in words.

---

3. You are mowing the lawn. As you mow, the amount of grass to be cut decreases. You mow at the same rate until about half the grass has been cut. Then you take a break for a while. Then, mowing at the same rate as before, you finish cutting the grass. Sketch a graph that shows how much uncut grass is left as you mow, take your break, and finish mowing.

From Graphs to Stories

Name ______________________________

1. John and his father participate in a 100-meter race. John started the race 3 seconds after his father began to run. The graph provides information about how far John and his father ran over time. Write a story about who won the race; be descriptive about how the race was run. If the two lines describing how each person ran were parallel, what would the graph tell you about who won the race?

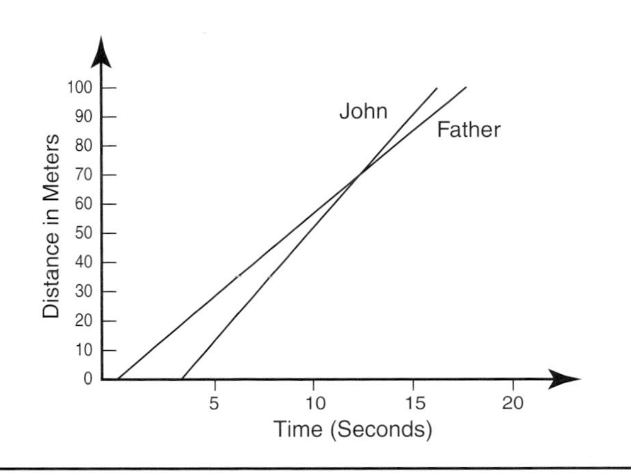

2. The graph represents the relationship between the profit and the amount of lemonade sold at a lemonade stand. Write a story about how the lemonade stand's profit is determined. Include an explanation of what is indicated when the line is below zero and when the line crosses the horizontal axis. (This graph assumes that the seller is not paid and that there is no overhead.)

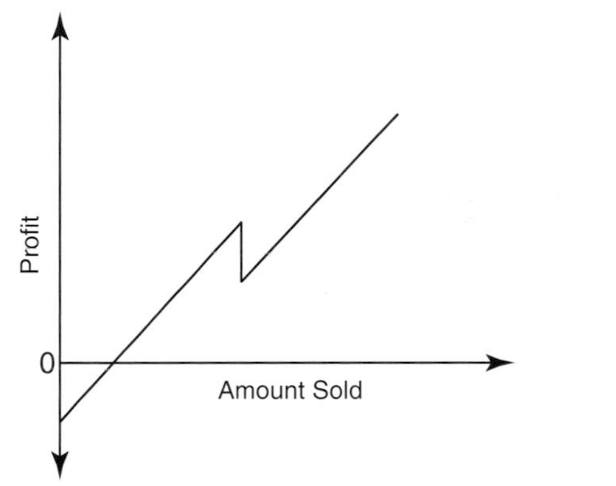

3. The graph represents a flag being raised on a flagpole. Write a story that describes what is happening to the flag, gives an estimate of the height of the flagpole, and explains the shape of the graph.

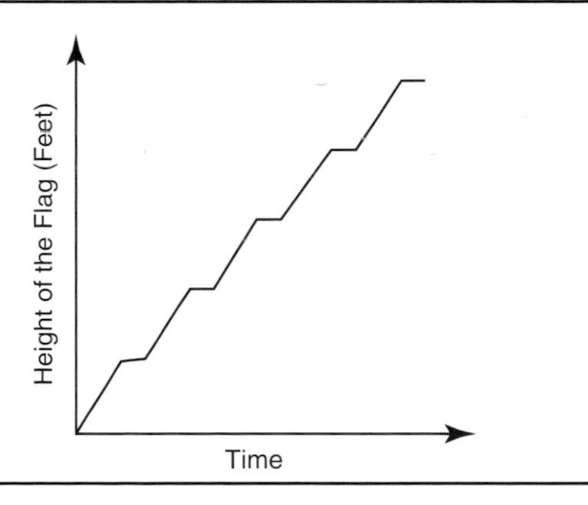

4. Graphs can be used to depict the story of a race. Here is a graph that represents a swimming race that occurred between two middle-grades students. Write a story that describes what happened in this race.

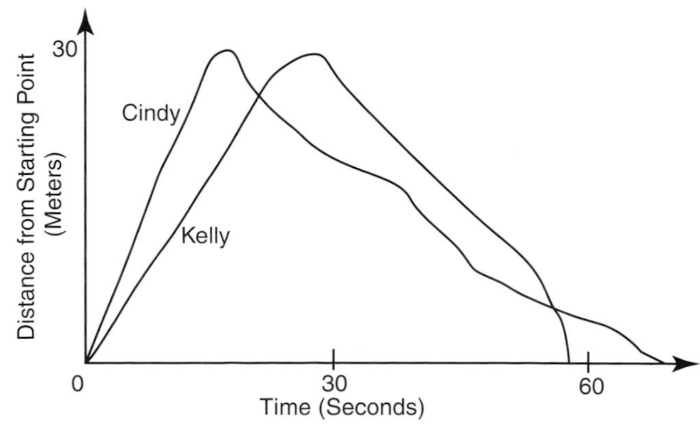

Missing Values

Name ______________________________

A piece of paper with an incomplete table showing number pairs was left on a student's desk after a mathematics class. Using what you know about patterns, answer these questions:

x	y
2	____
3	____
4	16
5	20
6	24
7	____
8	32

1. Predict the missing values for y, and record them in the table. Describe how you made your predictions.

   ____

   ____

2. Graph the data in the table on the coordinate grid below.

3. Describe a general rule to help you determine the value for y if you are given the value for x.

   ____

4. Use your rule to find the value for y when $x = -1$. Show this point on the line you graphed.

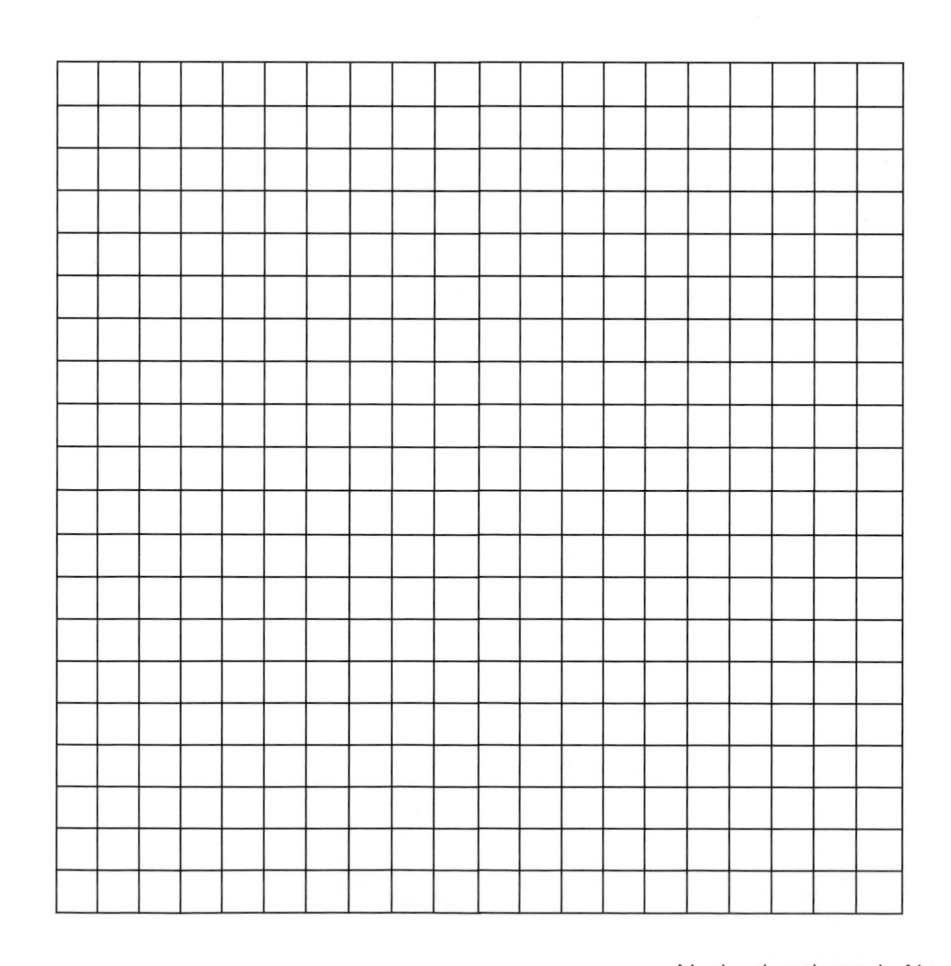

Stacking Cups

Name ______________________________

You have been hired by a company that makes all kinds of cups—foam hot cups, plastic drinking cups, paper cups, and more—of different sizes. For each of the kinds of cups it makes, the company needs to know the measurements of cartons that can hold 50 cups. Your task is to provide this information.

1. Make a table and record in it the measurement data (number of cups and height of a stack of cups) for different kinds of cups.
2. Make a coordinate graph of each data set.
3. Describe what variables are being investigated and the relationship between the variables.

   ____

   ____

   ____

4. Predict how tall a stack of 50 cups would be and explain how you made your prediction.

   ____

   ____

   ____

5. For each kind of cup, recommend the inside dimensions of a carton that would hold a stack of 50 cups. ______________________________

   ____

   ____

6. Compare the results for the different kinds of cups, noting similarities and differences.

   ____

   ____

   ____

Walking Rates

Name ______________________________

Several students are planning to participate in a walkathon to help raise money for charity. The distance to be walked is 10 kilometers. The students are wondering how long it might take them to walk this distance. Several students decide to do an experiment to determine their walking rates. Here are data for three of the students:

Name	Walking Rate
Jeff	1 meter per second
Rachel	1.5 meters per second
Annie	2 meters per second

From these walking rates, a table can be organized that shows the distance walked by each of the students after various numbers of seconds.

Time (Seconds)	Distance (Meters)		
	Jeff	Rachel	Annie
0	0	0	0
1	1	1.5	2
2			
3			

1. Complete the table for several times.
2. On one coordinate grid, make a graph of the time and distance data for the three people. Describe how the walking rates affect the graphs. ______________________________

   ____

   ____

3. For each of the three people, describe the relationship between the time and the distance walked, using words. Write an equation for each relationship, using d to represent distance in meters and t to represent time in seconds. Describe how the walking rate affects the equation.

   ____

   ____

   ______________________________.

4. Using these equations, determine the distance traveled in 1 minute, 30 minutes, and 1 hour. Use this information to estimate how much time it will take for each person to complete the walkathon. ______________________________

   ____

   ____

Pledge Plans

Name ______________________________

Several students who are participating in a 10-kilometer walkathon to raise money for charity need to decide on a plan for sponsors to pledge money for the walkathon. Jeff thinks that $1.50 per kilometer would be an appropriate pledge. Rachel suggests $2.50 per kilometer because it would bring in more money. Annie says that if they ask for too much money, people won't agree to be sponsors; she suggests that they ask for a donation of $4.00 and then $0.75 per kilometer.

1. Make a table showing the amount of money a sponsor would owe under each of the pledge plans.
2. On one coordinate grid, make a graph for each of the three pledge plans. Use a dashed line to connect the points.
3. For each of the three pledge plans, use words to describe the relationship between the money earned and the distance walked. ______________________________

____

____

____

____

4. Using m to represent the money owed, write an equation that can be used to compute the money owed under each pledge plan, given the distance the student walks. Use d to represent distance. ______________________________
5. Describe how increasing the amount of the pledge per kilometer affects the table, the graphs, and the equations.______________________________

____

____

____

____

6. Describe what is different about Annie's plan. What happens in the table, the graph, and the equation when Annie's plan is introduced? ______________________________

____

____

____

____

Fund Raising

Name ______________________________

The Middle Grades Math Club needs to raise money for its spring trip to Washington, D.C., and for its Valentine's Day dance. The club members decided to have a car wash to raise money for these projects. The club treasurer provided the following information:

- The total cost of sponges, rags, soap, buckets, and other materials needed will be $117.50.
- The usual charge for washing one car is $4.00.

As a member of the club, your job is to produce a mathematical analysis of this situation that includes making a table and a graph to answer the following questions:

1. How much money would the club raise if the members washed 0, 1, 2, 3, or 4 cars? __________

   ____________ 10 cars? ____________ 50 cars? ____________ 100 cars? ____________

   How can you use your table to answer these questions? ____________________

   How can you use your graph to answer these questions? ____________________

2. (*a*) If no cars are washed, what happens to the club's finances? How is this situation shown on the graph? ______________________ On the table? ____________________

   (*b*) With every car washed, what is the change in the amount of money earned? ____________
   How is this change shown on the graph? ____________ On the table? ____________

   (*c*) Write a general rule to use to determine how much money is raised for any number of cars. How is the information you identified for parts (*a*) and (*b*) related to the rule you wrote?

   ____

3. How many cars must the members wash to "break even"?____________________

4. How many cars must the members wash to raise enough money for—

   - The trip to Washington—estimated cost, $875?____________________
   - The Valentine's Day dance—estimated cost, $350? ____________________

5. Realistically, can the car wash raise enough money for both of these activities? ____________

   Explain why or why not. ____

   ____

   ____

   ____

Printing Books

Name ______________________________

Purchasing decisions for an algebra textbook for the school system must be made. The Board of Education needs to know the costs of providing a 325-page algebra textbook to three different schools. The schools have agreed to pilot a new textbook this year; the publisher is making page proofs of the book available, but the schools must make the copies needed for the students. An assistant has done some research and has discovered the following three possibilities:

1. A local printing company: The algebra textbook can be printed by a local printer for a cost of $9.50 per book with an initial cost of $5000 for typesetting.
2. A local copy center: The algebra textbook can be duplicated at a local copying center for $0.05 per page plus $2.00 per book for binding.
3. The school district: The school district's own copying center can reproduce the textbook at a cost of $0.035 per page plus an up-front cost of $3000.

In the past, the algebra textbook orders have never exceeded 2250 books and the cost has never exceeded $35,000. Using tables and graphs, as well as identifying general rules, slopes, y-intercepts, and points of intersection, do a mathematical analysis of these different options and write a recommendation for the Board of Education to consider.

Tiling Tubs

Name ______________________________

Hot tubs and in-ground swimming pools are sometimes surrounded by borders of tiles. This drawing shows a square hot tub with sides of length s feet. This tub is surrounded by a border of square tiles. Each border tile measures 1 foot on each side.

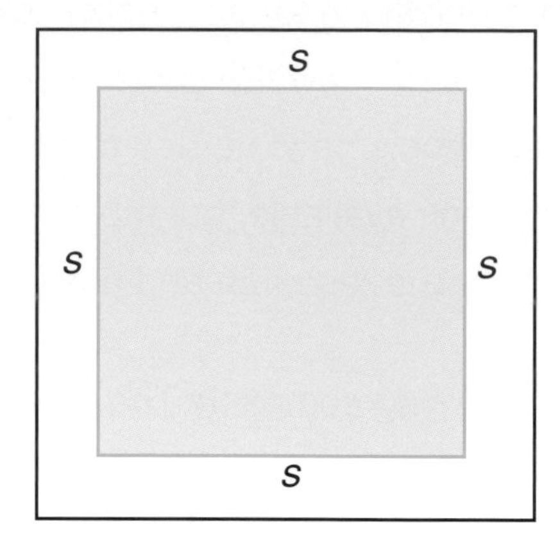

1. How many 1-foot square tiles will be needed for the border of a square hot tub that has edge length s feet? ______________________________

2. Express the total number of tiles in as many ways as you can. ______________________________

____

____

____

3. Be prepared to convince your classmates that the expressions are equivalent.

Plotting Land

Name ______________________________

A gardener has divided a plot of land into two sections. The shaded section is to be used for growing vegetables, and the unshaded portion is for growing flowers.

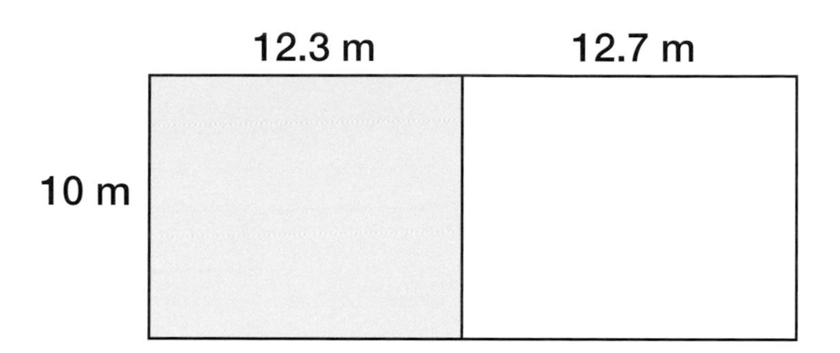

1. What is the area of the shaded portion of the plot of land? ______________________________

2. What is the area of the unshaded portion of the plot of land? ______________________________

3. What is the area of the entire plot of land? ______________________________

4. Suppose the gardener wanted to divide the plot of land so that 3/5 of the land was used for vegetables and 2/5 of the land was used for flowers.

   (*a*) What would be the dimensions of the shaded and unshaded portions? ______________________________

   (*b*) What would be the areas of the shaded and unshaded portions of the plot of land? ______________________________

   (*c*) What would be the area of the entire plot of land? ______________________________

References

Beckmann, Charlene E., and Kara Rozanski. "Graphs in Real Time." *Mathematics Teaching in the Middle School* 5 (October 1999): 92–99.

Billstein, Rick, and Jim Williamson. *MathThematics, Book 3*. Boston: McDougall Littell, 1999.

Bowman, Anita Hill. "Preservice Elementary Teachers' Performance on Tasks Involving Building, Interpreting, and Using Linear Mathematical Models Based on Scientific Data as a Function of Data Collection Activities." (Doctoral diss., University of North Carolina at Greensboro, 1993.) *Dissertation Abstracts International* 54 (1993): 3693.

Burkhardt, Hugh, Phil Daro, Jim Ridgway, Judah Schwartz, and Sandra Wilcox. *Middle Grades Assessment: Balanced Assessment for the Mathematics Curriculum, Package 2*. White Plains, N.Y.: Dale Seymour Publications, 2000.

Cai, Jinfa, and Patricia Ann Kenney. "Fostering Mathematical Thinking through Multiple Solutions." *Mathematics Teaching in the Middle School* 5 (April 2000): 534–39.

Cole, Beth R., and Gail Burrill. *Expressions and Formulas*. Chicago: Encyclopaedia Britannica, 1998.

Friel, Susan N. "Exploring How One Problem Contributes to Student Learning." *Mathematics Teaching in the Middle School* 4 (October 1998): 100–103.

Hadley, Bill. Cognitive Tutor Algebra I. Version 2.0. Pittsburgh: Carnegie Learning, 2000.

Hersberger, Jim, and Bill Frederick. "Flower Beds and Landscape Consultants: Making Connections in Middle School Mathematics." *Mathematics Teaching in the Middle School* 1 (April-May 1995): 364–67.

Johnson, Iris D. "Mission Possible! Can You Walk Your Talk?" *Mathematics Teaching in the Middle School* 6 (October 2000): 132–34.

Jones, Graham A., and Roger Day. *Algebra, Data, and Probability Explorations for Middle School: A Graphics Calculator Approach*. Menlo Park, Calif.: Dale Seymour Publications, 1998.

Kleiman, Glenn, Rosemary Caddy, Malcom Swan, Hugh Burkhardt, Shelley Isaacson, and Robert Bates. *Mathematics of Motion: Distance, Speed, and Time*. Teacher's ed. Mountain View, Calif.: Creative Publications, 1998.

Koirala, Hari P., and Phillip M. Goodwin. "Teaching Algebra in the Middle Grades Using Mathmagic." *Mathematics Teaching in the Middle School* 5 (May 2000): 563–66.

Lambdin, Diana V., R. Kathleen Lynch, and Heidi McDaniel. "Algebra in the Middle Grades." *Mathematics Teaching in the Middle School* 6 (November 2000): 195–98.

Lamon, Susan J. *Teaching Fractions and Ratios for Understanding*. Mahwah, N.J.: Lawrence Erlbaum Associates, 1999.

Lane, Suzanne. "The Conceptual Framework for the Development of a Mathematics Performance Assessment Instrument." *Educational Measurement Issues and Practice* 14, no. 1 (1993): 16–23.

Lapp, Douglas A., and Vivian Flora Cyrus. "Using Data-Collection Devices to Enhance Students' Understanding." *Mathematics Teacher* 93 (September 2000): 504–10.

Lappan, Glenda, James T. Fey, William M. Fitzgerald, Susan N. Friel, and Elizabeth Difanis Phillips. *Moving Straight Ahead*. Menlo Park, Calif.: Dale Seymour Publications, 1998a.

———. *Say It with Symbols*. Menlo Park, Calif.: Dale Seymour Publications, 1998b.

Mason, John, Alan Graham, David Pimm, and Norman Gowar. *Routes to/Roots of Algebra.* Milton Keynes, England: Open University Press, 1985.

National Council of Teacher of Mathematics (NCTM). *Principles and Standards for School Mathematics.* Reston, Va.: NCTM, 2000.

Nickerson, Susan D., Cherie Nydam, and Janet S. Bowers. "Linking Algebraic Concepts and Contexts: Every Picture Tells a Story." *Mathematics Teaching in the Middle School* 6 (October 2000): 92–98.

O'Connor, Peggy. "Uncovering the Magic." *Teaching Children Mathematics* 7 (September 2000): 16–21.

Rubenstein, Rheta N. "The Function Game." *Mathematics Teaching in the Middle School* 2 (November-December 1996): 74–78.

Silver, Edward A., and Suzanne Lane. "Assessment in the Context of Mathematics Instruction Reform: The Design of Assessment in the QUASAR Project." In *Cases of Assessment in Mathematics*, edited by M. Niss, pp. 59–69. London: Kluwer Academic Publishers, 1993.

TIMSS International Study Center. *Third International Mathematics and Science Study (TIMSS).* Boston: Boston College, TIMSS International Study Center, 1997.

Van de Walle, John A. *Elementary and Middle School Mathematics: Teaching Developmentally.* 4th ed. White Plains, N.Y.: Addison Wesley–Longman Publishing, 2000.

Suggested Reading

Day, Roger, and Graham A. Jones. "Building Bridges to Algebraic Thinking." *Mathematics Teaching in the Middle School* 2 (February 1997): 208–12.

Edwards, Thomas G. "Some 'Big Ideas' of Algebra in the Middle Grades." *Mathematics Teaching in the Middle School* 6 (September 2000): 26–31.

Fouche, Katheryn. "Algebra for Everyone: Start Early." *Mathematics Teaching in the Middle School* 2 (February 1997): 226–29.

McCoy, Leah P. "Algebra: Real-Life Investigations in a Lab Setting." *Mathematics Teaching in the Middle School* 2 (February 1997): 220–24.

Smith, John P. III, and Elizabeth A. Phillips. "Listening to Middle School Students' Algebraic Thinking." *Mathematics Teaching in the Middle School* 6 (November 2000): 156–61.

Stacey, Kaye, and Mollie MacGregor. "Building Foundations for Algebra." *Mathematics Teaching in the Middle School* 2 (February 1997): 252–60.